# Coal and Modern Coal Processing:
# An Introduction

# Coal and Modern Coal Processing: An Introduction

*Edited by*

## G. J. PITT
*Coal Research Establishment*
*National Coal Board, Cheltenham, England*

*and*

## G. R. MILLWARD
*Edward Davies Chemical Laboratories,*
*University College of Wales, Aberystwyth, Wales*

**1979**

ACADEMIC PRESS

London · New York · San Francisco

*A Subsidiary of Harcourt Brace Jovanovich, publishers*

ACADEMIC PRESS INC. (LONDON) LTD.
24/28 Oval Road,
London NW1

*United States Edition published by*
ACADEMIC PRESS INC.
11 Fifth Avenue
New York, New York 10003

Library of Congress Catalog Card Number 78-75271
Coal and modern coal processing.
1. Coal
I. Pitt, G J II. Millward, G R
662'.62        TN800        78-75271

ISBN 0-12-557850-4

Text set in 10/12 pt. Compugraphic English and printed in England at
The Lavenham Press, Lavenham, Suffolk

# Contributors

A. D. DAINTON    *Coal Research Establishment, National Coal Board, Stoke Orchard, Cheltenham, Glos.*

J. GIBSON    *National Coal Board, Stoke Orchard, Cheltenham, Glos.*

G. R. MILLWARD*    *Edward Davies Chemical Laboratories, The University College of Wales, Aberystwyth*

J. OWEN    *Coal Research Establishment, National Coal Board, Stoke Orchard, Cheltenham, Glos.*

G. J. PITT    *Coal Research Establishment, National Coal Board, Stoke Orchard, Cheltenham, Glos.*

*Now at Department of Physical Chemistry, University of Cambridge, Lensfield Road, Cambridge.*

# Foreword

One of life's ironies is that the action most needed for the sake of the future may be the least easy to implement in the present. The coal industry is an example. It has long been apparent that oil and gas supplies will soon become scarce and costly; that greater recourse will have to be had to coal as well as to nuclear power; and that, because of the amount of time needed to undertake all the research, development and investment required, there must be strong and immediate action if a breakdown in energy supplies to the consumer is to be avoided. The quadrupling of oil prices at the end of 1973 drove this lesson home to most people. Yet during the last five years the appropriate action has proved to be too difficult to undertake. The slowing down of economic growth and the ample supplies of hydro-carbons at present available as the result of investments already in hand have obscured the true situation and made it too expensive and unpopular to develop coal production and coal-based oil and gas supplies on the scale appropriate to the eventual need for them.

In this situation it is fortunate that we at least have gifted men who are continuing to develop coal science and technology, and that a number of them have co-operated to prepare this succinct and up-to-date introductory work on coal and coal processes. It is also appropriate that the book should have been based on lectures given in the Chemistry Department of the University College of Wales, Aberystwyth, to commemorate Dr Walter Idris Jones. An old student of the Department, he became the National Coal Board's first Director General of Research and a President of the Institute of Fuel, and he was keenly aware of the long-term energy problem and the part to be played by coal in its solution. He would have rejoiced to see the publication of such a useful book as this for the new generation of scientists and engineers.

SIR GORONWY DANIEL
*Vice Chancellor of the*
*University of Wales*

# Preface

During the 1976-77 session at the University College of Wales, Aberystwyth, the distinguished coal scientist Dr Walter Idris Jones was commemorated by a series of lectures on the science and technology of coal, arranged by Professor J. M. Thomas in the Edward Davies Chemical Laboratories. The lectures, designed to appeal to graduates and undergraduates in a wide range of disciplines, aimed to review knowledge on the nature and main properties of coal, and to discuss processes already established or under development for the conversion of coal into products such as coke, oils, synthesis gas and graphite, and for the generation of heat and electric power.

The eight lectures in the series form the basis for the present volume, supplemented by the insertion of Chapter 5 which describes the application of high resolution electron microscopy to the study of the structure of coal-derived materials. The latter work from the Edward Davies Chemical Laboratories at Aberystwyth provides valuable additional information to that presented in the lectures.

With the growing realization of the future importance of coal, increasing numbers of scientists and engineers are becoming involved in research and development work on new coal processes. Chapters 1 and 2 provide an outline account of the nature of coal, its classification, structure and special properties on which the newcomer to coal technology can build a more detailed knowledge on specific topics by consulting the available works of reference.

Coke, the main product at present manufactured from coal, forms the subject of Chapter 3, in which the reader can discern the relevance of many of the topics discussed in the first two chapters. The structures of cokes and graphites derived from coal are of importance to their performance in steelmaking, and various structural features are discussed in Chapter 4, mainly in the context of the materials science of these substances.

Chapters 6 to 8 deal with three basic types of process on which development is now in progress to meet the requirements for fuel, power and chemical feedstocks as reserves of oil decline. In the circumstances, no account of these topics can hope to be comprehensive or completely up-to-date; the aim has therefore been to indicate the main alternative lines along which work is being undertaken, together with some appraisal of their merits. Finally, Chapter 9 surveys the changing fortunes of by-products from the carbonization of coal and shows examples of how new materials, such as pitch polymer sheets, are being made from coal tar pitch, carbon fibres and graphite electrodes from coal extracts, and active carbon from anthracite.

Our thanks are due to numerous members of the Coal Research Establishment for their willing and efficient help in the preparation of the manuscript and the illustrations.

*March, 1979*                                                    G. J. PITT
                                                              G. R. MILLWARD

# Contents

# 1 Coal: an Introduction to its Formation and Properties

## J. GIBSON

*Board Member for Science, National Coal Board*

## I. INTRODUCTION

The element carbon plays such an overwhelmingly important role in life that its relative scarcity may come as a surprise. It accounts for only 0·04% of the total mass of the earth, and a mere 0·02% of the carbon in the earth's *crust* to a depth of 5 km occurs in a form which can react with oxygen, the rest being in the form of carbonates, carbon dioxide, etc. Most of this reactive carbon occurs in very concentrated form in the fossil fuels, peat, lignite, coal, oil and natural gas, and the known reserves of coal represent by far the largest proportion.

Fossil fuels are today not only the main source of energy but also the basis of many manufacturing processes (plastics, pharmaceutical products, iron and steel, aluminium and many other materials in daily use). Some of these products are currently manufactured by the use of carbon derived from oil, but there is almost unanimity in the opinion that the production of oil will peak towards the end of the century and then decline sharply, thereby increasing our dependence on coal for reactive carbon.

Coal is therefore a very important natural resource and much effort has been expended (and is still being expended) on making the best of it. This has been paralleled by an intensive investigation of coal structure, which has proved to be a challenging subject for scientific study. In this book both of these aspects of coal science—technological use and fundamental scientific study—will have their place, since they are intimately interwoven in present day research and development.

This chapter surveys briefly the formation of coal and provides some basic

1

information concerning its composition and classification, together with a little about its chemical and physical properties, which will form a background for the remaining chapters.

## II. THE FORMATION OF COAL

### Origins

Coal is not just another form of carbon, such as graphite or diamond. It consists of a complex mixture of organic chemical substances containing carbon, hydrogen and oxygen in chemical combination, together with smaller amounts of nitrogen and sulphur. Coal is, in fact, a fossil formed mainly by the action of temperature and pressure on plant debris. This organic part of coal has associated with it various amounts of moisture and minerals.

In certain parts of the world there existed, some 300 million years ago, warm and humid climatic conditions, which favoured the growth of huge tropical ferns and giant trees, which grew and died in vast swamp areas. The dead plants fell into the boggy waters, which tended tò exclude oxygen and kill bacteria, and here they only partially decomposed to produce a peat-like material. Fossilized plant remains can be found in coal measure rocks (Fig. 1a), and fossilized spores (Fig. 1b) have an important application in identifying coal seams in borehole samples.

Vegetation continued to grow for many generations, forming vast, thick peat beds which were later to turn into coal. However, after a time the areas of swamp gradually became submerged and were covered with sedimentary deposits. As the rate of subsidence of the coal basin was not constant, this cycle of swamp followed by submersion was often repeated a number of times, with the result that a sequence of horizontal bands of peat and inorganic, sedimentary rocks was built up. This formed the first stage, the biochemical stage. Subsequently, the bands of peat were altered by the action of pressure and temperature during the second, or geochemical stage to form the coals found today. As much as a 20-fold reduction in the thickness of the original plant deposit sometimes occurred.

During the course of time these horizontal coal seams were further altered as they became folded, tilted and/or eroded. This sequence of events is shown diagrammatically in Fig. 2.

### Coalification

Coalification is the name given to the development of the series of substances peat, lignite or brown coal, bituminous coal and, finally, anthracite. The degree of coalification or rank of the coal increases progressively from lignite

*Fig. 1a.* Neuropteris heterophylla *Brongniart from the Coal Measures near Barnsley, Yorks* (× 1). (*Reproduced by permission of the Director, Institute of Geological Sciences; NERC copyright.*)

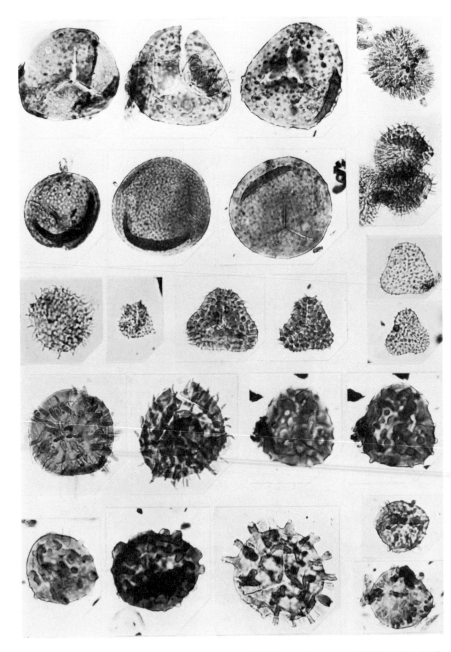

*Fig. 1b. Carboniferous miospores from British coal seams* (×360). (*Copied with the permission of the Palaeontological Association from Smith and Butterworth, 1967.*)

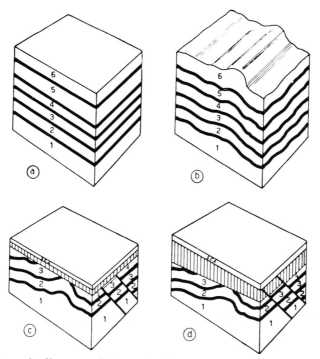

*Fig. 2. Tectonic diagram of the carboniferous strata (coal seams in black).
(a) Undisturbed horizontal strata, (b) after folding, (c) after erosion, faulting
and deposition of overburden, (d) after tilting of fault blocks (reproduced
with permission from van Krevelen, 1961).*

through low-rank coal to high-rank coal to anthracite.* The carbon content
increases and the oxygen and hydrogen contents decrease throughout the
series, while the reactivity decreases.

Differences in the plant material and the extent of its decomposition
during the first stage largely determined the different petrographic types
known as *macerals*, and the subsequent action of pressure and heat during
the geochemical stage caused the differences in coalification or maturity of
the coals, usually known as the *rank* of the coal. Macerals (so-called by
analogy with minerals in inorganic rocks) are commonly grouped together
under the names vitrinite, exinite and inertinite. They can be distinguished
by the use of microscopy, and their properties are found to be different,
although the differences become smaller as the rank increases. Figure 3 gives
an example of the convergence of properties at high rank.

Vitrinite is commonly the major petrographic constituent of British coals
and the most consistent in its properties. Many scientific investigations have

*Peat and lignite are comparatively rare in Britain and are excluded from the rest of
the book.

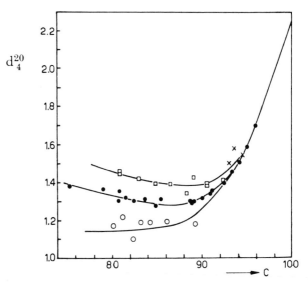

*Fig. 3. Densities of coal macerals.* • *Vitrinites,* o *Exinites,* □ *Micrinites,* x *Fusinites (reproduced with permission from van Krevelen, 1961).*

therefore concentrated mainly on the vitrinite which can often be selected in reasonable purity from large lumps of coal; this would appear to be preferable to working with a series of whole coals, which will differ among themselves in maceral group composition. The other macerals, however, must not be ignored. Some coals—notably from the Southern Hemisphere—contain less vitrinite and their properties are in some respects markedly different from those of Britain and other Northern Hemisphere countries.

Coal, therefore, is a term which embraces a family of substances covering a range of ranks, each substance containing a number of macerals in varying proportions (not to mention moisture and inorganic mineral matter). It is clearly important to be able to analyse and classify a given coal in such a way that its properties can be predicted.

*Table I. Chemical compositions of wood, peat and various coals.*

|  | % C | % H | % O | % N |
|---|---|---|---|---|
| Wood | 50·0 | 6·3 | 42·7 | 1·0 |
| Peat | 57·0 | 5·2 | 36·8 | 1·0 |
| Lignite | 65·0 | 4·0 | 30·0 | 1·0 |
| Low-rank coal | 79·0 | 5·5 | 14·0 | 1·5 |
| Medium-rank coal | 88·0 | 5·3 | 5·0 | 1·7 |
| Anthracite | 94·0 | 2·9 | 1·9 | 1·2 |

## Composition

Coal can be analysed chemically to determine the percentages of the main elements present, namely carbon, hydrogen, oxygen and nitrogen. Typical analyses of coals of different ranks are given in Table I, with comparative values for wood, peat and lignite. These analyses are expressed on a dry, mineral-matter-free basis to avoid the situation where two batches of a coal would have different analyses because they contained different amounts of moisture or mineral matter. (This is not the only complication which arises as oxygen and sulphur can occur in both the coal and the mineral matter, but these are problems for the analyst.) It is sufficient to note (as shown in Fig. 4) that the vitrinites of low-rank coal contain up to about 15% oxygen and 5% hydrogen, and that with increasing rank the oxygen decreases first with little change in the hydrogen content, but later the hydrogen content decreases too. Note also, that exinites contain more hydrogen and less oxygen than the corresponding vitrinites, whereas the reverse is true for micrinite (one form of inertinite).

Diagrams of the above type provide an overall view of the coalification process. Another method of representing the chemical analyses of coals and their precursors is given in Fig. 5 in which H/C is plotted against O/C; from such diagrams it appears that during coalification of wood, the early stages involved dehydration followed by decarboxylation in the middle stages and demethanation later.

## III. STRUCTURE OF COAL

### Classification

Chemical analyses of the kind just discussed can clearly provide a basis for classifying coals, and one way of doing this was proposed nearly 80 years ago by a famous British coal scientist, Dr Seyler. He based his scheme on the carbon and hydrogen contents (which were much easier to determine than the oxygen content) and he found that coals tended to fall into a well-defined band when their composition was plotted, as shown in Fig. 6.

A number of the technological properties of coals, such as the calorific value, can be read off by the use of appropriate grids superimposed on the Seyler chart. The coal band in this chart is, of course, closely related to Figs 4 and 5 which illustrate that, with increasing coal rank, the oxygen content decreases first with little change in the hydrogen content (and consequently there is an increase in carbon content), and in the later stages of coalification the hydrogen content decreases.

While this type of classification had much to recommend its use in scientific studies, it is not used in commercial practice because here it is more convenient to base classification on properties of more direct importance in

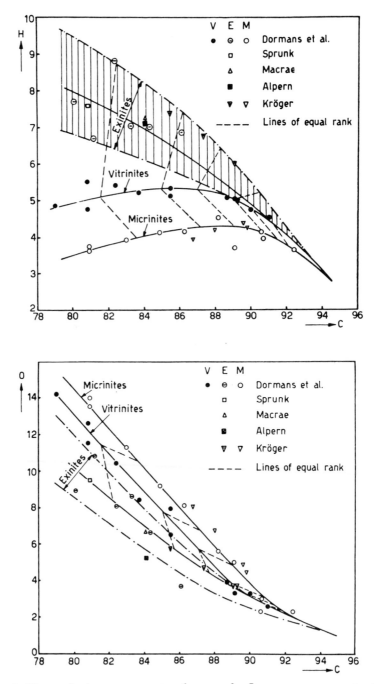

Fig. 4. Upper: hydrogen contents of macerals. Lower: oxygen contents of macerals (reproduced with permission from van Krevelen (1961)).

coal processing. These properties can be determined by less skilled operators using methods which are closely standardized in any given country. International acceptance of a common standard classification may eventually be achieved, but Britain still adheres to its own system (as do some other countries).

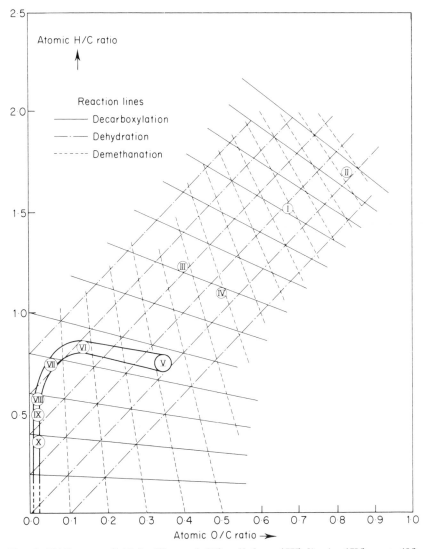

*Fig. 5. H/C versus O/C for (I) wood, (II) cellulose, (III) lignin, (IV) peat, (V) lignite, (VI) low-rank coal, (VII) medium-rank coal, (VIII) high-rank coal, (IX) semi-anthracite, (X) anthracite (reproduced from van Krevelen (1950) by permission of the publishers, IPC Business Press Ltd. (C)).*

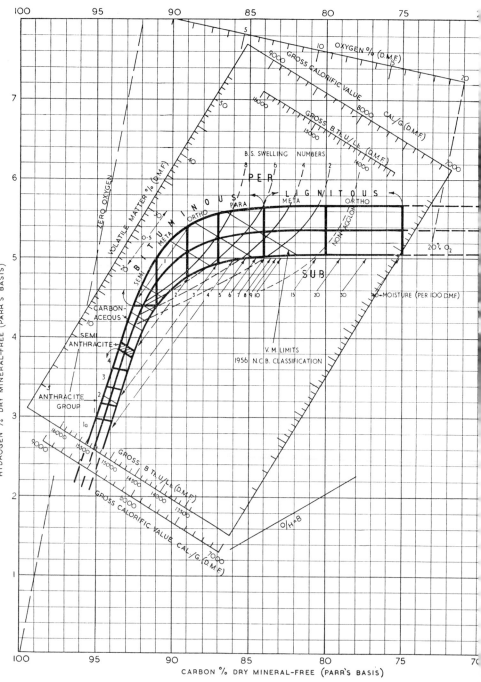

*Fig. 6. Seyler's coal chart.*

In most countries the classification is primarily based on the so-called volatile matter content. This term, unfortunately, suggests that coals are a mixture of components, some of which can be distilled off, whereas others cannot. In fact, it means the percentage loss in weight when a sample of crushed coal is rapidly heated to 900°C under standard conditions in a crucible of a specified type, allowance being made for moisture and mineral matter in the coal. Volatile matter is therefore a measure of the amount of material driven off under standardized decomposition conditions rather than distillation of pre-existing components; it decreases with increasing rank from about 40% to about 5%. In a given coal, exinite has a higher volatile matter content than vitrinite, whereas inertinite has a lower value.

The use of volatile matter as a parameter in coal classification has obvious relevance to the carbonization of coal, giving a rough measure of the yields of tar plus gas which may be expected when a coal is heated in the absence of oxygen. Its acceptance over the whole field of coal processing is probably explained by its wide and fairly uniform variation from low-rank to high-rank and its ease of measurement. (One unfortunate feature of the use of volatile matter as a basis for classification is that high-volatile corresponds to low-rank and vice versa, causing occasional confusion.)

While volatile matter forms the primary basis of coal classification, it is found necessary in practice to use at least one more parameter in order to characterize the behaviour of a coal when heated, because two coals may have the same volatile content but differ from one another in the extent to which they swell up or cake together. In Britain, the parameter is known as the Gray-King coke type and is based on the appearance of a coke specimen prepared under standard conditions (Fig. 7). The details of this parameter need not concern us here, but Fig. 8 shows its use in the current NCB classification scheme, particularly to sub-divide high-volatile coals into six different degrees of caking capacity.

The rank series of British coals starts with high-volatile, non-caking coals suitable for combustion (and hence electricity generation), progresses through coals of increasing caking capacity, which are of value as components of blends for carbonization (see Chapter 3), to prime coking coals which make excellent coke, but are now in very short supply. It then progresses to steam coals, which are also used as components of blends for carbonization and are good fuels for stoves since they produce little or no smoke, and finally to anthracites, noted for smokeless combustion and used by the NCB as a starting material from which active carbon can be made. There are other uses for coals, some of which will feature in later chapters but enough has been said to convey an idea of the diversity which exists. Figure 9 illustrates the variations in the coalfield of South Wales, where there is a regular progression from east to west, from medium caking coals through prime coking coals to steam coals and anthracites. A similar spread of rank is found in the coalfields of the Eastern USA and to a lesser extent in the coalfields of France and Germany.

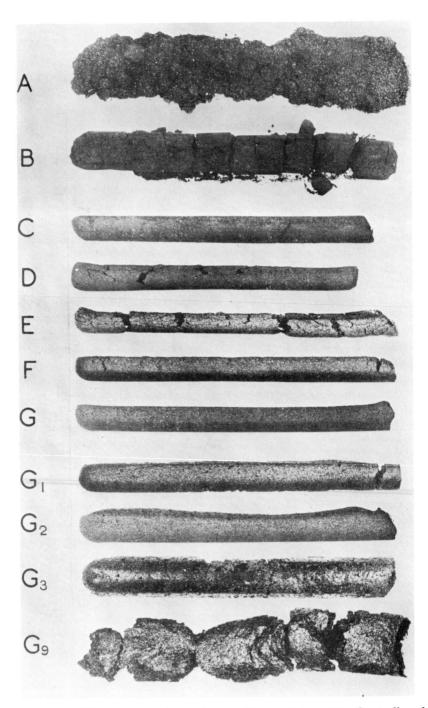

Fig. 7. Gray-King coke types (reproduced with permission of the Controller of HMSO from 'The Efficient Use of Fuel' (1958)).

Before leaving the subject of coal classification, it should be pointed out that the Seyler classification scheme and the commercial schemes depend on the measurement of properties made on a sample of powdered coal. The values obtained are therefore averages over a large number of particles and if the batch of coal happens to be a mixture (from two different seams in a colliery for example) it is not apparent that this is so. Furthermore, if two batches of the same coal differ markedly in the proportions of the different maceral groups, they may not be recognizable from the measured parameters

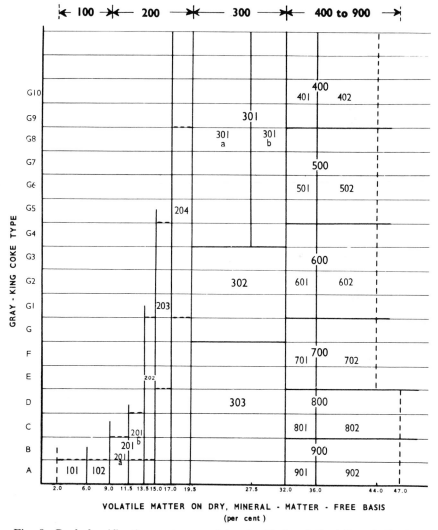

Fig. 8. *Coal classification system used by the National Coal Board. Broken line defines a genera limit as found in practice, although not a boundary for classification purposes; solid line defines a classification boundary.*

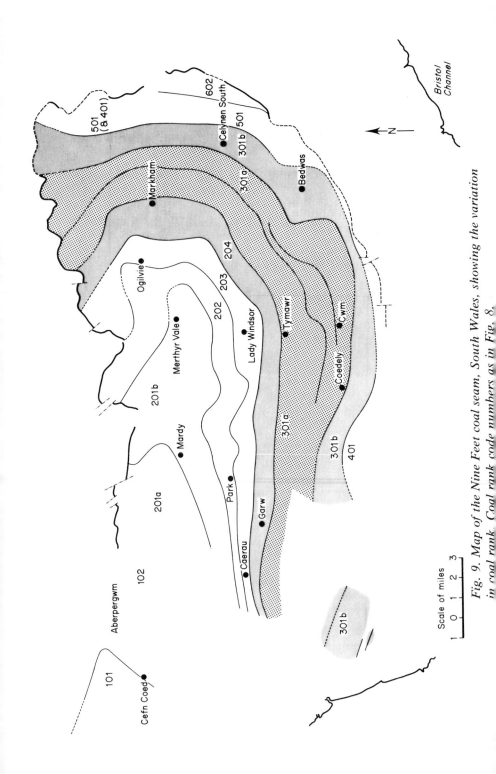

Fig. 9. Map of the Nine Feet coal seam, South Wales, showing the variation in coal rank. Coal rank code numbers as in Fig. 8.

as coming from the same source. Coal petrography, which is the microscopic study of coal as a rock, presents the possibility of overcoming these analytical difficulties, but only at a greatly increased cost. This is one of the topics discussed in Chapter 2.

## Molecular structure

The chemical analysis of coal having been established, it must be admitted that a knowledge of the carbon, hydrogen and oxygen contents does not convey much about its chemical make-up. Even in elementary organic chemistry, a substance of molecular weight 46 containing 52% carbon, 13% hydrogen and 35% oxygen might be ethanol or dimethylether, and if the molecular weight is unknown, the number of possible alternatives becomes boundless. Other evidence is necessary before it is possible to define the way in which the various atoms are assembled, and investigations into the structure of coal are described in Chapter 2. In the meantime, it will suffice to say that a coal is not a single chemical compound with identical molecules, but that it can be regarded as a statistical structure made up of small, condensed, aromatic units or layers with a variety of substituent groups around the perimeter and some cross-linking between adjacent units. Some of the units may not be strictly planar because of the presence of hetero-atoms (oxygen, nitrogen or sulphur) or hydroaromatic portions, but they approximate to planarity and consequently show some tendency to pack parallel to one another, although this parallelism is not very extensive. The structure of graphite consists of extensive parallel layers of carbon atoms arranged in condensed, aromatic arrays, the layers being mutually orientated and separated by a distance of 0·335 nm (the carbon-carbon van der Waals' distance). The parallel stacking of layers in coal is similar, but lacks mutual orientation between layers, and the average spacing between layers is somewhat larger.

The three main types of coal, namely low-rank (non-coking or weakly caking), medium-rank (coking) and high-rank correspond approximately to three types of structure (Fig. 10). The first type, characteristic of low-rank coals, has small layers more or less randomly orientated and connected by cross-links, so that the structure is highly porous. The second type, characteristic of coking coals, shows a greater degree of orientation and consequently a greater tendency towards parallel stacking; there are less cross-links, fewer pores and the structure has some similarities to that of a liquid. The third type, seen in high-rank coals and anthracites, shows a growth of the individual layers, a marked increase in the degree of orientation and the development of a new type of pore elongated parallel to the stacks of layers. The progressive changes throughout the coalification process can thus be visualized.

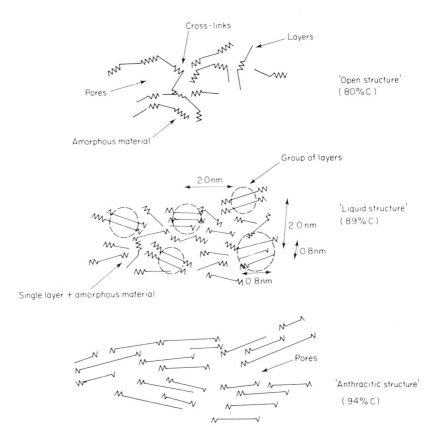

*Fig. 10. Structure types for a low-rank coal, a coking coal and an anthracite (reproduced with permission from Hirsch, 1954).*

## IV. THE PROPERTIES OF COAL

### Behaviour on heating

From a practical standpoint the most important properties of coal are those associated with its behaviour when heated, with or without the presence of air, since carbonization accounts for about 20% of the coal currently used in Britain, and virtually all the rest is burned in one way or another.

When heated, most coals evolve tarry vapours, gas and moisture (Fig. 11) and some coals soften and fuse into a coke residue. In the presence of air, the combustible products burn. The tarry vapours normally burn with a smoky flame, the gases burn with a non-smoky flame and the solid residue glows, leaving ash derived from the mineral matter. The calorific value of coal varies with rank in a manner which can be predicted from the changes in chemical composition, so that low-rank coals generate less heat because of their higher oxygen content (and because of their higher moisture content), but they are widely burned because of greater availability and lower cost. If the combustion conditions lead to the production of smoke, there is not only a pollution problem, but also a reduction in efficiency, and these are important considerations behind the work described in Chapter 6.

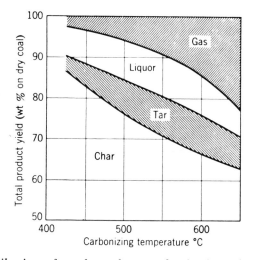

*Fig. 11. Distribution of products from carbonization of coal at various temperatures (reproduced by courtesy of the Institute of Fuel from Owen, 1958).*

In the absence of air, the sequence of changes which occur when coal is heated forms the basis of the carbonization process from which metallurgical coke, domestic coke, coal tar, town's gas (now replaced by natural gas) and other products, including ammonia, are derived. This subject is discussed in more detail later, but it is appropriate to point out here that the way coals behave when heated is consistent with the coal structural types described earlier. In the first place, the tar is derived from aromatic layer "molecules" in the original structure which are only weakly held by cross-links and can therefore be evolved as a result of minor decomposition. This process is limited because the breakage of cross-links generates free valencies and these tend to satisfy themselves by recombination. Gaseous products such as methane and hydrogen are derived from the breakage of bonds to peripheral substituent groups and combination of the resulting radicals; this also leaves free valencies and increases the incidence of recombination so that the coke remaining becomes more cross-linked and involatile.

When heated, coking coals soften in the temperature range 400-500°C, become plastic and agglomerate. It would appear that plasticity depends on the availability of an adequate concentration of liquid decomposition products and a limited degree of cross-linking of the network of layers in the coal. There is less cross-linking in a coking coal than in a low-rank coal, and plasticization is therefore enhanced. In high-rank coals on the other hand, there is a lower concentration of liquid decomposition products and therefore plasticization decreases again. Evidently there is a delicate balance and it can be shifted to some extent. For example, increasing the rate of heating leads to increasing evidence of plasticity in a low-rank coal.

## The treatment of coal with solvents

The use of coal in any process involves solids-handling techniques, and for some time considerable attention has been devoted to the possibility of dissolving coal in solvents so that liquid phase technology can be used. Many years earlier, coal chemists made use of solvent extraction as a technique for fractionating coal. These early studies were based on the view that coal consisted of a mixture of organic substances, including a so-called "coking principle" which was responsible for the good coking properties of certain coals. It was hoped that poor coking coals might be upgraded by addition of the coking principle obtained by solvent extraction. This hope was not fulfilled in commercial terms, but the work led to the investigation of the action of a wide range of solvents on coals and established a connection between the action of solvents and the phenomena observed in carbonization.

Conventional solvents for coal can be grouped into categories as follows:

(1) Non-specific solvents, which extract not more than a few per cent of a coal at temperatures up to about 100°C, the extract being considered to come from resins and waxes which do not form a major part of the coal substance. Ethanol is an example.

(2) Specific solvents, which extract 20-40% of the coal at temperatures below 200°C, the most effective ones being nucleophilic and having electron-donor capacities. The nature of the extract is closely similar to that of the original coal. Pyridine is an example of this class of solvent.

(3) Degrading solvents, which extract major amounts of the coal (up to more than 90%) at temperatures up to 400°C; the solvent can be recovered substantially unchanged from the solution. Their action is presumed to depend on a mild thermal degradation of the coal to give smaller and more soluble fragments. Phenanthrene is an example.

(4) Reactive solvents, which dissolve coal by reacting with it. They are mostly hydrogen donors (for example, tetralin) and their chemical com-

position is appreciably affected during the process; as a consequence of this reaction the extracts differ markedly in properties from those obtained with degrading solvents.

In recent years considerable attention has been given to the use of compressed gases as solvents in extraction processes. Application of Dalton's law of partial pressures to a gas in contact with a substance of fairly low volatility would suggest that, in most conditions, the concentration of

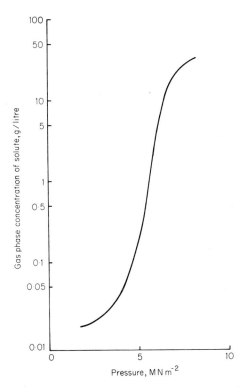

*Fig. 12. Solubility of* p-*iodochlorobenzene in ethylene at 15°C as a function of pressure* (*reproduced by courtesy of the Institute of Fuel from Whitehead and Williams, 1975*).

"solute" in the gas phase would be very low and would decrease as the pressure was increased. However, very large deviations from Dalton's law occur at temperatures near the critical temperature of the gas and the concentration of solute in the gas is then much enhanced, and, moreover, *increases* with pressure. Figure 12 shows that the 'solubility' of *p*-iodo-chlorobenzene in ethylene can be increased a thousand times in this way.

The importance of choosing a solvent gas with critical temperature close to the extraction temperature is shown in Fig. 13 where carbon dioxide, ethane and ethylene are effective solvents at 40°C, but gases with lower critical temperatures are not.

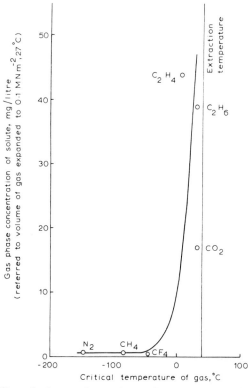

*Fig. 13. Solubility of phenanthrene in various gases at 40°C, 40 MNm⁻²* (*reproduced by courtesy of the Institute of Fuel from Whitehead and Williams, 1975*).

The ability to extract relatively involatile substances is particularly useful for those materials which decompose before reaching boiling point. The technique is therefore well suited to the extraction of the liquids formed when coal is heated to about 400°C. These liquids are only partially distillable, the remainder normally being converted into polymerized, involatile residues and gas when coal is carbonized. Gas extraction affords a means of recovering the liquid products when they are first formed, avoiding undesirable secondary reactions. Yields of extract up to 25 or 30% are thus possible.

Although these yields are lower than can be obtained with some liquid solvents and the use of high pressures might seem disadvantageous, gas extraction has a number of attractive features:

(1) Gas extracts have lower molecular weights (about 500 compared to greater than 2000) and higher hydrogen contents and may therefore be more readily converted to useful products.

(2) Solvent removal and recovery is simpler and more complete; reduction of the pressure precipitates the extract almost completely.

Current research on the liquefaction of coal, in which solvents of the anthracene oil type and compressed gases such as toluene are both being used, is described in Chapter 8.

## Porosity

The porous structure of coal has been studied in some instances to supplement knowledge of coal structure and in others for its potential commercial exploitation. The production of active carbons for commercial use is described in Chapter 9 and only the fundamental aspects will be mentioned here.

The fraction of the volume of a solid occupied by pores can be determined from measurements when the solid is immersed first in one fluid which fills the pores completely and then in another fluid which does not enter the pores at all. Helium and mercury are commonly chosen as the two immersion media. The density of coal in helium, being that of the pore-free solid, changes with rank (Fig. 14) because of changes in chemical composition and aromaticity. Combining these observations with those on the corresponding densities in mercury, we find that the graph of porosity against a rank parameter shows a minimum for the coking coals (Fig. 15).

The size distribution of the pores can be investigated by immersing the coal in mercury and progressively increasing the pressure on the mercury. Because of surface tension the mercury cannot enter pores with a diameter smaller than a value $d$ for any given value of pressure $p$, where $p = \dfrac{4\sigma \cos \Theta}{d}$

$\sigma$ being the surface tension and $\Theta$ the angle of contact. By recording the amount of mercury entering the coal for each small increment of pressure, it is possible to build up a picture of the distribution of sizes of pores (Fig. 16). It is found that the total pore volume accounted for is significantly less than that calculated from the helium density, and it has therefore been concluded that coal contains two pore systems: a macropore system accessible to mercury under pressure and a micropore system inaccessible to mercury but accessible to helium. By using liquids of various molecular sizes it is possible to investigate the distribution of micropore sizes, and a bimodal distribution is found (Fig. 17) with peaks corresponding to diameters of about 0·5 and

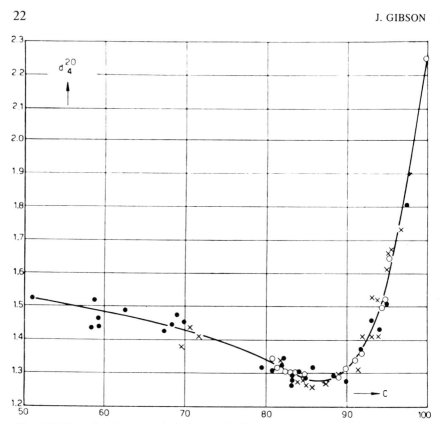

*Fig. 14. True densities of vitrinites:* • *Dulhunty and Penrose;* o *Franklin;* x *Zwietering (reproduced with permission from van Krevelen, 1961).*

0·8 nm. Although the relationship of these micropores to the structural models of coal is as yet not fully understood, it must involve the interstices between the "molecules", the coal behaving in some respects like a molecular sieve.

In a number of contexts the presence of a large internal surface area is of greater significance than the actual pore volume, for instance, in the adsorption of substances on active carbon (see Chapter 9), and in the mining of coal where sorbed methane is encountered.

## V. LITERATURE

Later chapters deal more fully with the topics introduced above and if the reader wishes to explore the subject further the following books are recommended:

(1) Francis, W. (1961). "Coal: Its Formation and Composition." (2nd ed.) Edward Arnold (Publishers) Ltd., London.

(2) van Krevelen, D. W. (1961). "Coal. Typology-chemistry-physics-constitution." Elsevier, Amsterdam.

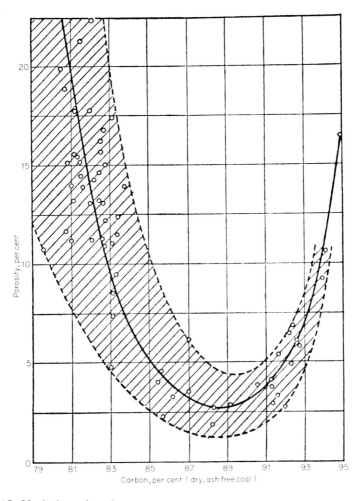

*Fig. 15. Variation of coal porosity with rank (King and Wilkins, 1944).*

(3) Lowry, H. H. (Ed.) (1945). "Chemistry of Coal Utilisation." (2 vols, supplementary vol., 1963.) John Wiley, New York. (At present being completely revised but sufficiently up to date for the general reader.)

(4) Monographs on chemical engineering published by Mills and Boon, London:

     (a) D. G. Skinner, (1971). "Fluidised Combustion of Coal."

     (b) Gibson, J. and Gregory, D. H. (1971). "Carbonisation of Coal."

     (c) Wise, W. S. (1971). "Solvent Treatment of Coal."

     (d) Paul, P. F. M. and Wise, W. S. (1971). "Principles of Gas Extraction."

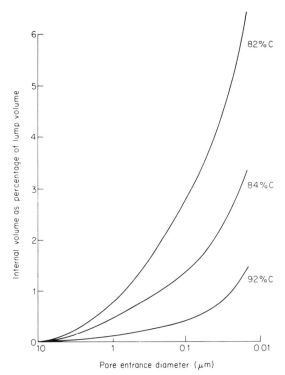

Fig. 16. Distribution of sizes of pores in coals of various ranks.

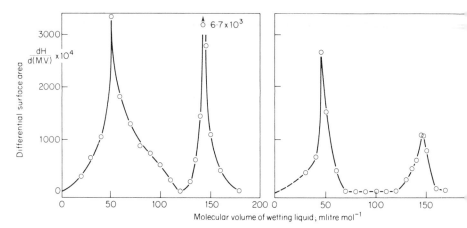

Fig. 17. Distribution of surface area with pore size for (left) low-rank coal, (right) high-rank coal (reproduced with permission from van Krevelen, 1961).

The first two are readable accounts of fundamental coal science written shortly after the period of great activity in this field in the 1950s, in which Prof. van Krevelen and his school played an important role. The earlier volumes of Lowry's book contain much information which is now largely historical, but the supplementary volume contains useful accounts of some of the more modern developments in coal processing and a condensed account of the fundamental work. Finally, the Mills and Boon series provides recent concise accounts of new techniques in course of exploitation at present. Unfortunately, all of these books except the supplementary volume of "Chemistry of Coal Utilisation" are at present (April 1978) out of print.

## REFERENCES

Hirsch, P. B. (1954). *Proc. R. Soc.* A **226,** 143—169.
King, J. G. and Wilkins, E. T. (1944). *In* "Proc. Conf. Ultra-fine Structure of Coals and Cokes." pp. 46-56. B.C.U.R.A., London.
Owen, J. (1958). *In* "Residential Conf. on Science in the Use of Coal." pp. C-34-39. Inst. of Fuel, London.
Smith, A. H. V. and Butterworth, M. A. (1967). "Spec. Papers in Palaeontology No. 1." The Palaeontological Assoc., London.
van Krevelen, D. W. (1950). *Fuel, Lond.* **29,** 269-284.
van Krevelen, D. W. (1961). "Coal. Typology-Chemistry-Physics-Constitution." Elsevier, Amsterdam.
Whitehead, J. C. and Williams, D. F. (1975). *J. Inst. Fuel* **48,** 182-4.

# 2  Structural Analysis of Coal

## G. J. PITT

*Coal Research Establishment, National Coal Board*

## I. INTRODUCTION

In the first chapter, a brief account of the composition and 'molecular structure' of coal was given and this topic will now be dealt with in greater detail. Coal is by no means a simple substance and a reliable impression of its structure can only be achieved by a synthesis of results from many different methods of investigation such as X-ray analysis, infra-red spectroscopy, NMR and the use of specific chemical reagents. The structure of coal attracted considerable attention, particular in the 1950s, and although notable advances have been made, it must be emphasized that many aspects of the structure are still a subject for discussion and research.

## II. OPTICAL METHODS OF ANALYSIS

It will be helpful to begin by considering what can be observed when coal is examined under the microscope. A photomicrograph of a very thin section (Plate 1 facing p. 74) shows that coal is a heterogeneous substance containing several different entities, called macerals, which are grouped together under the names vitrinite, exinite and inertinite. These have been found to differ from one another in composition and properties and so it is clear that they should be studied separately to obtain as much information as possible. Vitrinite is the predominant maceral group in many coals and can be obtained in relatively pure state by hand selection and density separation. Many of the investigations of coal structure have therefore given most attention to the vitrinite, so the exinite and inertinite (which can be concentrated less readily) have been studied to a lesser degree.

For general microscopic study of coal, thin sections are too difficult to prepare and would be impossible to make from crushed coal. The normal practice is therefore to mount the coal particles in a block of epoxy resin and then polish a flat surface through the block. In the sections of the particles exposed, the macerals can be distinguished (Fig. 1) and their concentrations estimated by a counting procedure. In addition to this another important

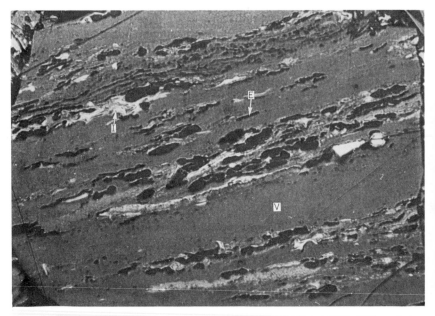

*Fig. 1. Photomicrograph of polished coal section. V: vitrinite; E: exinite; I: inertinite.*

property can be studied, namely the reflectance of the vitrinite, which increases with the rank of the coal. This property can be measured by mounting a photomultiplier on a microscope so that it receives light reflected from a very small area on the polished surface (about 2 microns in diameter), and comparing the intensity with that reflected from the same area on a standard surface of known reflectance. Since there is always a variation from point to point even within a coal from a single seam, measurements are usually made on 100 to 500 such areas of vitrinite to obtain the distribution of values (Fig. 2). In a given coal, the reflectances of the exinite and inertinite are lower and higher respectively than that of the vitrinite, but they are measured only for special purposes (Fig. 3).

Reflectance is of importance in the utilization of coal because it is the only parameter of rank which can be measured on individual particles of coal (in

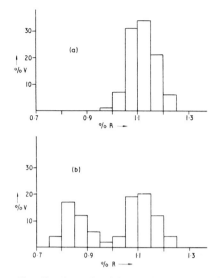

*Fig. 2. Reflectance distributions for (a) single seam coal (b) blend of two coals (reproduced with permission from Juckes and Pitt, 1977).*

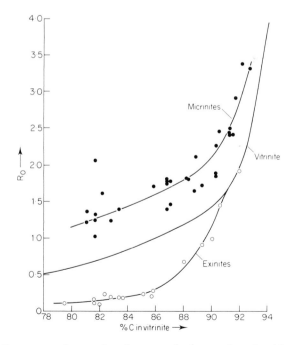

*Fig. 3. Reflectance of associated macerals (reproduced with permission from van Krevelen, 1961).*

contrast to carbon content or volatile matter which are bulk properties) and it can therefore be used to discover whether a sample of crushed coal is a mixture or not (Fig. 2). It is also the only measurable parameter of rank which is not influenced by changes in the proportions of the maceral groups. In combination with the concentrations of the maceral groups it can be used to predict the characteristics of coke made from a coal and the yields of products. For the investigation of coal structure, on the other hand, reflectance is of use because it is related to the refractive index and hence the molar refraction, which is an additive property of the atoms present and the bonds between them; this is referred to later in this chapter. The reflectance of most coals varies with direction relative to the bedding plane of the deposit and this is of interest since it shows the influence of pressure during the coalification process. The degree of anisotropy is negligible in low-rank coals, increases progressively through medium-rank coals and becomes quite marked in anthracites.

## III. SKELETAL MOLECULAR STRUCTURE

### X-ray analysis

The discussion of the application of microscopy to the study of coal leads naturally to the question how the structure at the 'molecular' level can be investigated. In the analysis of crystalline substances, X-ray diffraction is usually capable of establishing the molecular arrangement, even in extremely complicated molecules such as haemoglobin, because of the high degree of regularity in the crystal lattice. Coal, however, is not a crystalline substance and the molecular structure cannot be deduced from X-ray analysis; there is, indeed, no evidence to suggest that there is a coal 'molecule' in the strict sense, although it is useful to retain the concept on a loose basis.

The chemical structure of coal involves a carbon skeleton, and X-ray analysis has provided very useful information on the arrangement of the carbon atoms. The diffraction patterns derived from coals are very diffuse but examination shows that there are broad peaks present in positions related to those of the sharp peaks given by crystalline graphite (Fig. 4).

Broadened peaks are commonly associated with small crystallite size but this is not the complete explanation, because a sequence of graphite peaks such as 100, 101, 102, 103 is replaced by one broad, asymmetrical peak. This phenomenon also occurs in other carbonaceous substances such as carbon black, and it can be explained if there are small layer planes present of the graphite type, but they are stacked without regular order from one to the next. Two questions then arise: (1) how large are the layers and (2) how many layers are there in a parallel stack? These can best be answered by fitting the

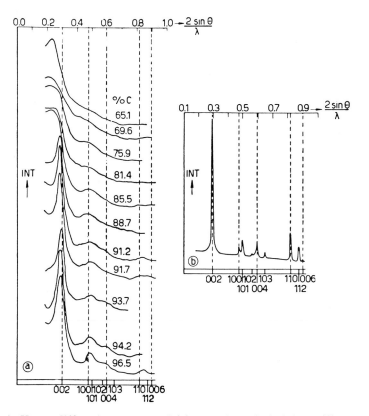

*Fig. 4. X-ray diffraction curves of (a) a series of vitrinites, (b) graphite (reproduced with permission from van Krevelen, 1961).*

appropriate part of the X-ray intensity curve using a distribution of (1) layer sizes and (2) stack heights.

The distribution of layer sizes can be calculated on the assumption that layers fall into a number of classes (Fig. 5). More than one arrangement of atoms is taken for each class in order to average out the characteristics specific to any one molecule. It is found (Fig. 6) that the smallest size class (mean diameter 0·6 nm) occurs with the highest frequency in vitrinites of all ranks from 78 to 94%C, and classes with mean diameters greater than 1·0 nm are scarcely observed except in the coals of highest rank. There is an appreciable content of amorphous material in all coals, the amount decreasing progressively as rank increases. The average layer diameter changes little up to about 90%C, but thereafter increases rapidly (Fig. 7). It must be emphasized that there are various ways in which this form of analysis could give a false impression. What is registered by X-rays is the regular array of scattering centres on a planar hexagonal network, and anything which interrupts the regularity (holes, deviations from planarity, etc.) effectively terminates the layer, whilst anything which extends the regularity (substituent atoms in approximately the correct position) will enlarge the apparent size of the layer. It would therefore be very unwise to attach strictly quantitative significance to the histograms in Fig. 6.

The distribution of heights of the parallel stacks of layers can be obtained by the application of a suitable mathematical transformation to the 002 peak in the intensity curve. Analyses of a low-rank coal and a coking coal (Fig. 8) show that many layers occur singly and that stacks of more than 3 layers are rare; the layers are better stacked in the coking coal than in the low-rank coal.

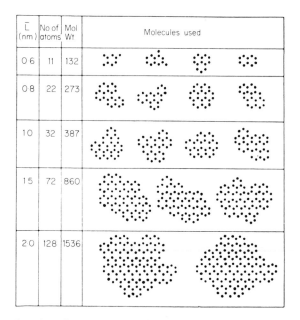

| $\bar{L}$ (nm) | No of atoms | Mol Wt | Molecules used |
|---|---|---|---|
| 0 6 | 11 | 132 | |
| 0 8 | 22 | 273 | |
| 1 0 | 32 | 387 | |
| 1 5 | 72 | 860 | |
| 2 0 | 128 | 1536 | |

*Fig. 5. Molecular size classes used in the interpretation of X-ray diffraction curves of coals and cokes (reproduced by permission of the International Union of Crystallography from Diamond, 1957).*

There are other features in the X-ray diffraction patterns of coals which should be mentioned. At very low angles of diffraction a marked diffuse scattering is observed which is attributed to the presence of porosity. The intensity of this scattering is a minimum for coking coals, as would be expected from the results of other methods of measuring porosity. In addition, medium and high-rank coals show a broad band with a spacing of 2 nm indicating a preferred distance of this magnitude between scattering centres, which may be a consequence of the increased degree of parallel stacking of layers.

The results of X-rays analysis of vitrinites may therefore be summarized in terms of three types of structure (Chapter 1, Fig. 10): (1) in low-rank coals, an open structure of small, condensed aromatic layers randomly orientated and cross-linked; (2) in medium-rank coals, a 'liquid' structure with fewer cross-links, a moderate degree of orientation and reduced porosity; (3) in

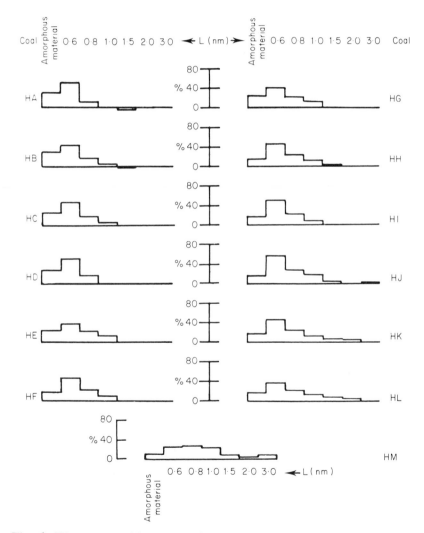

*Fig. 6. Histograms of layer sizes for a series of vitrains of different rank. Carbon content increases progressively from HA (78%) to HM (94%) (reproduced by courtesy of the Institute of Fuel from Hirsch, 1958).*

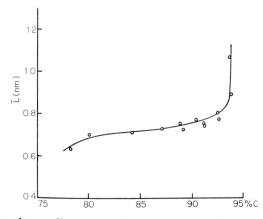

*Fig. 7. **Mean layer diameters of vitrains** of various carbon contents (reproduced by courtesy of the Institute of Fuel from Hirsch, 1958).*

high-rank coals, a structure of larger layers with a higher degree of orientation and an orientated pore system.

Exinites differ from vitrinites in showing a so-called γ band with spacing 0·4 to 0·5 nm which is thought to arise from non-aromatic groupings such as alicyclic layers or aliphatic side-chains. Inertinites give diffraction patterns similar to vitrinites but the layers appear to be larger and better stacked.

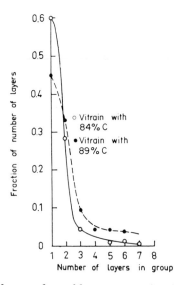

*Fig. 8. Fraction of the number of layers occurring in groups of n layers for vitrain of 84 and 89%C (reproduced with permission from Hirsch, 1954).*

## Statistical constitution analysis

It may be considered questionable that a few ill-defined bands in the X-ray diffraction pattern provide sufficient basis for the construction of a com- paratively well-defined picture of the carbon skeleton. There have been claims, for example, that some of the intensity distribution curves can be fitted equally well by alicyclic structures as by aromatics. In an attempt to obtain confirmatory evidence of the X-ray structure, a number of methods, generally described as statistical constitution analysis, have been employed which make use of the fact that various properties of compounds are additive, meaning that the value of the property can be calculated by summation of the contributions for the individual atoms, together with contributions associated with the types of bonds linking the atoms.

The simplest example of an additive property is molar volume which is equal to the sum of the single-bonded atomic volumes per gram atom minus a term to allow for multiple bonds. Using generally accepted coefficients we have:

$$\text{Molar volume} = \frac{M}{d} = 9{\cdot}9C + 3{\cdot}1H + 3{\cdot}8\,O + \ldots - K_M$$

where $M$ = molecular weight, $d$ = specific gravity; C, H, O. . . are numbers of carbon, hydrogen, oxygen . . . atoms per molecule and $K_M$ is a correction term for multiple bonds. Dividing by C and remembering that % carbon/100 = $12\,C/M$:

$$\frac{1200}{(\%C).d} = 9{\cdot}9 + 3{\cdot}1\,\frac{H}{C} + 3{\cdot}8\,\frac{O}{C}\ldots -(\text{term for aromatic C} = \text{C bonds}).$$

An equation of this form has been found to be satisfied by data derived from eighteen different polymers. By the use of a number of alkyl aromatic compounds and pitch fractions as reference substances, a calibration has been effected in terms of $f_a$, the fraction of the carbon in aromatic groups (Fig. 9). (Terms in O/C, N/C, etc. have been removed by applying corrections, which are small in practice.) Other properties can be used in a similar way including heat of combustion and sound velocity, giving $f_a$ values in reason- able agreement (Table I). There is still, however, some disagreement between different investigators over the correct values of $f_a$ and the available evidence can best be summarized by saying that a coal of 80%C has $f_a$ between 0·7 and 0·8 and a coal of 90% has $f_a$ between 0·8 and 0·9. NMR techniques may narrow these ranges in the near future, but existing methods are sufficient to confirm the highly aromatic nature of coal.

For confirmation of the size of the aromatic units deduced from X-ray

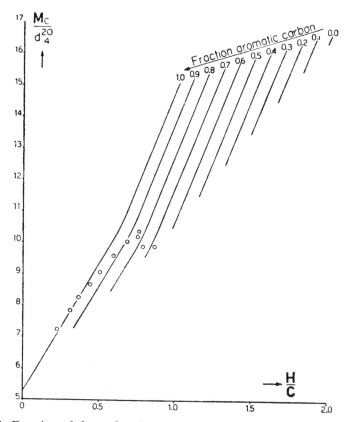

*Fig. 9. Fraction of the carbon in aromatic groups for a series of coals as a function of the molar volume per gram atom of carbon and the H/C ratio (reproduced with permission from van Krevelen and Schuyer, 1957).*

analysis, another additive property has been used, namely the molar refraction $\dfrac{(n^2 - 1)}{(n^2 + 2)} \cdot \dfrac{M}{d}$ where $n$ is the refractive index. If the reflectance $R$ of a coal is measured using two media with different refractive indices ($n_0$) in contact with the coal, the refractive index $n$ can be derived by solving two simultaneous equations

$$R = \frac{(n - n_0)^2 + k^2}{(n + n_0)^2 + k^2}$$

where $k$ is the (unknown) absorption coefficient of the coal (Fig. 10). Adopting the same procedure to eliminate the molecular weight, dividing by C gives:

$$\frac{n^2 - 1}{n^2 + 2} \cdot \frac{1200}{\%C.d} = 2{\cdot}6 + 1{\cdot}0 \frac{H}{C} + \ldots + I_m$$

*Table I. The aromaticity of vitrinites.*

| Carbon content % | Molar volume | Fractional aromaticity based on: | | |
| --- | --- | --- | --- | --- |
| | | Sound velocity | Heat of combustion | U.V. spectrum |
| 70·5 | 0·70 | | | |
| 75·5 | 0·78 | | | |
| 80·0 | | | ∿0·82 | |
| 81·5 | 0·83 | 0·79 | | |
| 84·0 | | | | >0·78 |
| 85·0 | 0·85 | 0·82 | ∿0·82 | |
| 87·0 | 0·86 | | | |
| 89·0 | 0·88 | 0·85 | | |
| 90·0 | 0·90 | | ∿0·85 | |
| 91·2 | 0·92 | 0·90 | | |
| 92·5 | 0·96 | 0·93 | | |
| 93·4 | 0·99 | 0·97 | | |
| 94·2 | 1·00 | | | |
| 95·0 | 1·00 | | 1·00 | |
| 96·0 | 1·00 | | | |

(Reproduced with permission from van Krevelen, 1961)

where $I_m$, the difference between the experimentally determined molar refraction and the value calculated from the atomic contributions, is assumed to be due to the aromatic C = C bonds. It can be shown that $I_m$ is theoretically proportional to the number of aromatic carbon atoms per

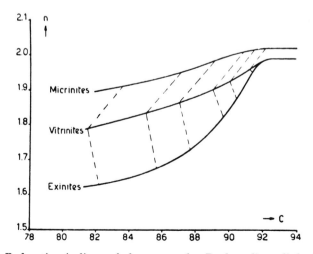

*Fig. 10. Refractive indices of the macerals. Broken lines link the points corresponding to the associated macerals in a given coal (reproduced with permission from van Krevelen, 1961).*

molecule ($C_a$) and is a function of the surface area of the molecule (this arises from consideration of the polarizability associated with mobile π-electrons in an aromatic skeleton). Experimental points for a number of pure compounds show satisfactory agreement with the theoretical relationship (Fig. 11), and there is some justification for applying the method to obtain estimates of the surface area of the aromatic layers in coal, from which can be calculated the number of carbon atoms in the aromatic layers. Table II shows that there is reasonable agreement with the values derived from X-ray analysis for coals of up to 87%C. For higher rank coals the values fall again, possibly because of the development of charge-transfer characteristics and semiconducting properties.

Table II also shows supporting evidence on the size of the aromatic layers obtained in a number of ways, using for instance the elementary composition (very high-rank coals contain few hydrogen atoms to saturate the edge valencies of aromatic molecules), semiconduction (the energy barrier Δ $u$ is a function of the number of aromatic carbon atoms per molecule in condensed aromatic compounds) and the UV spectrum (which can be synthesized from those of model compounds).

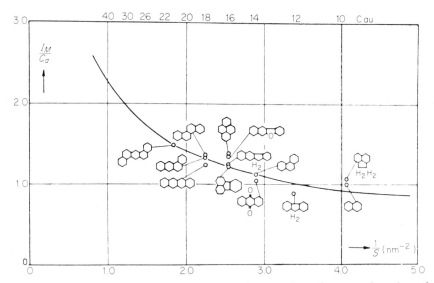

Fig. 11. Molar increment per gram atom of aromatic carbon as a function of aromatic surface area. Points from experimental data on model substances are shown for comparison with the theoretical curve (reproduced from Schuyer and van Krevelen, 1954, by permission of the publishers, IPC Business Press Ltd. (C)).

*Table II. Average numbers of carbon atoms per layer in vitrinites.*

| Carbon content % | X-ray diffraction | Elementary composition | Molar refraction increment | Semi-conductivity | U.V. spectrum |
|---|---|---|---|---|---|
| 70·5 | | | 12 | | |
| 75·5 | | | 13 | | |
| 81·5 | 16 | | 17 | | |
| 84·0 | 17 | | | | >17 |
| 85·0 | 17 | | 21 | | |
| 87·0 | 17 | | (23) | | |
| 89·0 | 18 | | | | |
| 90·0 | 18 | | | | |
| 91·2 | 18 | | | | |
| 92·5 | 18 | | | | |
| 93·4 | 20 | (20) | | (45) | |
| 94·2 | (30) | 22-40 | | (50) | |
| 95·0 | | 43-60 | | 55 | |
| 96·0 | | 85-100 | | > 60 | |

(Reproduced with permission from van Krevelen, 1961)

## IV. SIDE-GROUPS AND INTERMOLECULAR BONDING

With the carbon skeleton reasonably well defined, attention turns to the arrangement of the hydrogen and oxygen (and nitrogen, sulphur. . .) atoms. For information on the distribution of the hydrogen atoms, infra-red spectrophotometry is useful. As coal is a solid, it is necessary to use a suspension of finely divided coal in liquid paraffin or other mulling agent or in a potassium bromide disc as a specimen. Portions of the absorption spectrum of the coal are obscured by the suspension medium, but the spectrum can be mapped over an adequate range of frequencies by the use of both liquid paraffin and hexachlorobutadiene, or KBr (Fig. 12). The spectrum shows a number of distinct bands, the frequencies of which can be assigned with some confidence to particular groupings such as —OH, —$CH_2$, —$CH_3$, —CO and aromatic —CH. It is perhaps in the nature of coal that one of the largest peaks (1600 cm$^{-1}$) has been the subject of prolonged discussion over its assignment to aromatic C = C, chelated quinone groups or heterocyclic nitrogen. The absorption bands at 3030 cm$^{-1}$ and 2920 cm$^{-1}$ can be used to compare the amount of hydrogen attached to aromatic carbon and aliphatic carbon respectively. Since, for each peak, the absorbance is equal to the concentration of absorbing groups times the appropriate extinction coefficient we can deduce the ratio of the numbers of aromatic and aliphatic CH bonds if the ratio of the extinction coefficients is known. The last ratio has been estimated for a wide range of alkyl aromatic compounds,

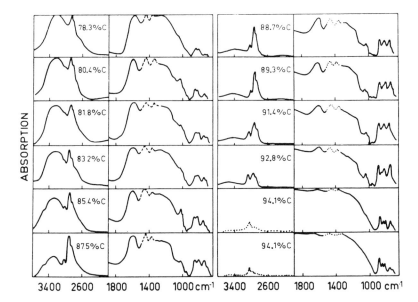

*Fig. 12. Infra-red spectra of a series of vitrains. 4000 to 2000 cm⁻¹ measured on hexachlorobutadiene mulls, 2000 to 650 cm⁻¹ on Nujol mulls (reproduced with permission from Brown, 1955).*

giving values between 0·3 and 1·0 with many near 0·5. Using this value, it is calculated that about four-fifths of the hydrogen in low-rank coals is aliphatic and that the proportion falls progressively with increasing rank becoming zero for an anthracite (Table III). As for other methods of structural analysis, there are a number of assumptions made, not all of which are generally accepted, but there is at least reasonable confirmation of the results from other techniques.

*Table III. The proportion of aromatic hydrogen in vitrinites.*

| Carbon content (%) | Infra-red spectrum | NMR spectrum | Composition of carbonization residue |
|---|---|---|---|
| 76·0 | | | 0·23 |
| 79·0 | | 0·33 | 0·33 |
| 83·0 | 0·18 | 0·45 | 0·38 |
| 88·0 | 0·28 | | 0·54 |
| 89·0 | 0·31 | | 0·51 |
| 91·0 | 0·45 | | |
| 93·0 | 0·64 | 0·64 | |
| 94·0 | 1·0 | | |

(Reproduced with permission from van Krevelen, 1961)

The first of these is nuclear magnetic resonance. Coal, being a solid, gives only a broad-line, proton spin resonance spectrum which is of limited usefulness, but the second moment of the absorption band depends on the relative proportions of different types of carbon-hydrogen bonds present, making it possible to derive an estimate of the proportion of aromatic hydrogen. The lack of detail in the coal spectrum can be partially overcome by the use of extracts in solution, which give spectra with peaks attributed to aromatic protons and benzylic protons and a group of peaks from hydrogen attached to saturated carbon atoms more than one $CH_2$ group removed from aromatic carbons (Fig. 13). Fewer assumptions are necessary in the evaluation of the proportion of aromatic hydrogen and other parameters from such spectra, but there is no guarantee that the extract has the same structure as the whole coal since it usually represents less than 20% of the coal. However, the results are in satisfactory agreement with those deduced from the broad line spectra.

The use of $^{13}C$ resonances instead of those due to protons presents the possibility of obtaining additional information relating to the structure of extracts (Fig. 14), but the applications of this technique so far reported have mainly been confined to the confirmation of aromaticities derived from proton resonances. Newer apparatus now becoming available increases the signal-to-noise ratio considerably and may therefore prove much more informative; cross-polarization techniques are also being developed for obtaining greater detail in the spectrum of coal itself which will enable the (carbon) aromaticity of coal to be measured directly. These techniques could yield information of considerable value in the near future.

Another estimate of the hydrogen aromaticity is shown in Table III, based on the chemical analyses of the coal and the residue after thermal decomposition, making assumptions about the fate of the aliphatic hydrogen and the free valencies occurring during the decomposition.

Some further information concerning the nature of the aliphatic hydrogen is also available from the infra-red spectra, which show that methylene groups are 6-25 times as numerous as methyl groups. Furthermore, reaction of coals with various agents known to dehydrogenate hydroaromatic systems (for example, triphenylmethyl perchlorate in boiling acetic acid) leads to the result that 25 to 40% of the hydrogen is removed. Bearing in mind that not all hydroaromatic systems are reducible, this indicates that much of the aliphatic part of coal is in fact hydroaromatic.

In contrast to the mainly physical methods used in assessing the groupings in which carbon and hydrogen occur, the methods employed to assess oxygen groupings are mainly chemical. Many different methods have been employed; there are 14 methods listed in one review of the subject for hydroxyl groups, for example, and 5 for carbonyl groups. The accessibility of groups in solid coal to chemical reagents is problematic and some investigators have

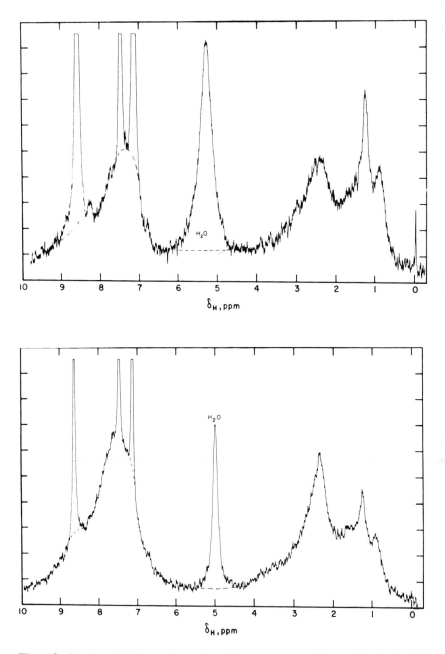

*Fig. 13. Proton NMR spectra of pyridine extracts of (upper) vitrain-rich Pittsburgh coal (lower) vitrain from Pocohontas No. 4 coal. The three sharp peaks at 7-9 ppm are due to pyridine (reproduced with permission from Retcofsky and Friedel, 1970).*

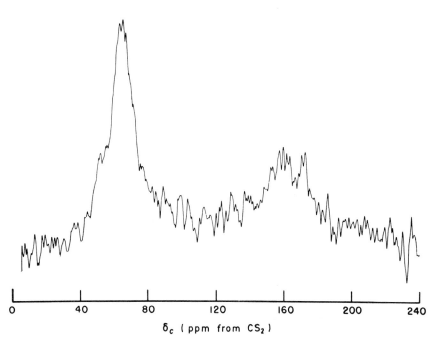

*Fig. 14. ¹³C NMR spectrum of carbon disulphide extract of Lower Banner coal (reproduced from Retcofsky and Friedel, 1976, by permission of the publishers, IPC Business Press Ltd. (C)).*

consequently used solvent extracts of coal, arguing that their infra-red spectra are sufficiently similar to those of the coal to justify this procedure. Although, relatively speaking, agreement between different methods is rather poor, it is probable that the majority of the oxygen in bituminous coals is in carbonyl and phenolic hydroxyl groups (in roughly equal proportions), the remainder being in ether and carboxyl groups or heterocyclic ring systems (Fig. 15).

Very little is known about the state of combination of the nitrogen and sulphur atoms in coal. From the presence of heterocyclic compounds in coal tar and the partial retention of both elements in coke it is likely that a proportion of the nitrogen and sulphur occurs in condensed aromatic structures, but some may be present as —NH, —SH or S—S groups.

## V. TOTAL MOLECULAR STRUCTURE

It is now of interest to attempt to assemble a structure from the carbon skeleton and the substituent groupings derived above. It might seem that this could be achieved in a vast number of ways, and in a sense it can. However, many alternatives are eliminated in practice because of the shortage of hydrogen (and especially of aromatic hydrogen) relative to carbon, the smallness of the aromatic nuclei and the steric problems of introducing appreciable cross-linking. Figure 16 shows an attempt to fit the known characteristics for a low-rank coal (80%C). There are a number of aromatic units of 1 or 2 benzene ring size in accordance with X-ray analysis, nitrogen

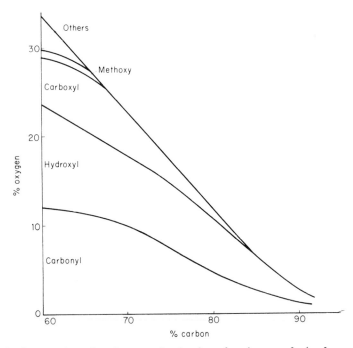

*Fig. 15. Oxygen functional groups in vitrains of various ranks (redrawn, after Blom, 1960).*

atoms replacing a few CH groups. These units (layers) are linked mainly by pairs of methylene bridges as in 9,10-dihydrophenanthrene, which are non-planar so they terminate the layers in X-ray diffraction. Because of these folded rings, the 'molecule' as a whole is far from flat, which explains the very limited degree of parallel stacking of layers that can occur, and also accounts for the fact that a solvent which swells the coal can extract a larger proportion of it. Clearly, there is likely to be considerable ultra-fine porosity in an assemblage of molecules of this sort.

It may be assumed that there is a distribution of 'molecules' with different molecular weights present in a tangled state in coal. The molecular weight chosen for the model is arbitrary, but within the range reported for solvent

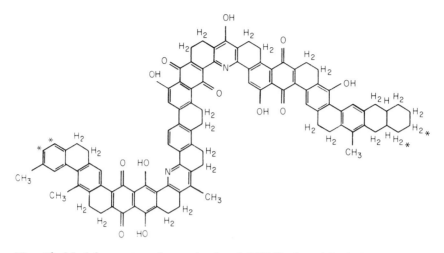

*Fig. 16. Model structure for a vitrain of 80%C. Asterisks indicate where dimerization could occur.*

extracts of low-rank coals. Solvents will extract a fraction of lower molecular weight so that the whole coal will probably have a higher average molecular weight corresponding to more extensive cross-linking or dimerization, etc.

The oxygen groupings shown provide approximately the right amounts of $C=O$ and OH, and, in addition, are an attempt to account for the strong $1600 \text{ cm}^{-1}$ band in the infra-red spectrum. Chelated quinones are one of the few types of carbonyl groups which absorb at a frequency as low as this, but there is no certain evidence of their occurrence in coal.

The choice of the 9,10-dihydrophenanthrene type of linkage between the aromatic units is based on evidence from three sources. Nuclear magnetic resonance studies on extracts and vacuum distillates show only a very small fraction of $CH_2$ groups directly linked to *two* aromatic nuclei. Secondly, oxidation studies suggest that there should be a significant number of aromatic nuclei joined by a single bond. Thirdly, dehydrogenation during pyrolysis leads to the formation of condensed aromatic nuclei, many of which are angular (like phenanthrene) rather than acenes (such as anthracene).

Figure 17 shows a possible 'molecule' for a higher rank coal of 90%C, in which the aromatic layers contain between 2 and 4 condensed benzene rings. The proportions of hydrogen and oxygen relative to carbon have become smaller but several characteristics of the lower rank coal (such as the 9,10-dihydrophenanthrene bridging) remain unchanged. Table IV shows the comparison between the properties of these model structures and coals of similar rank. Both models contain slightly more hydrogen than the corresponding coals, but this difference would become smaller if the model were

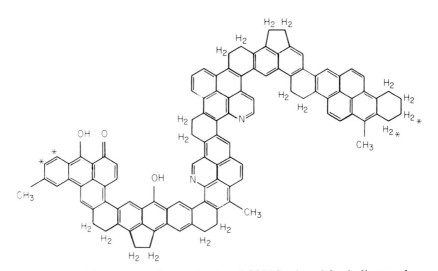

*Fig. 17. Model structure for a vitrain of 90%C. Asterisks indicate where dimerization could occur.*

*Table IV. Comparison between properties of model structures and coals.*

| Property | Model, Fig. 16 | Coal 80%C | Model, Fig. 17 | Coal 90%C |
|---|---|---|---|---|
| Chemical analysis | | | | |
| % carbon | 80·1 | 80·0 | 89·8 | 90·0 |
| hydrogen | 5·2 | 5·0 | 5·0 | 4·8 |
| nitrogen | 1·9 | 2·0[a] | 1·9 | 2·0 [a] |
| oxygen | 12·8 | 13·0 | 3·3 | 3·2 |
| Empirical formula | $C_{100} H_{78} N_2 O_{12}$ | | $C_{108} H_{72} N_2 O_3$ | |
| Molecular weight | 1498 | | 1444 | |
| Aromaticity ($f_a$) | 0·70 | 0·7 to 0·8 | 0·81 | 0·8 to 0·9 |
| Average number of atoms/layer | 13 | 14 | 20 | 18 |
| Aromatic CH/ aliphatic CH | 0·16 | 0·1 | 0·56 | 0·5 |
| Methylene groups/ methyl groups | 6 | ⩾6 | 6 | ⩾6 |
| Hydroxyl oxygen (% of coal) | 6·4 | 7·0 | 2·2 | 1·8 |
| Carbonyl oxygen (% of coal) | 6·4 | 5·1[b] | 1·1 | 1·4 |

(a) includes sulphur
(b) includes carboxyl oxygen

increased in molecular weight by dimerization. The agreement is otherwise satisfactory, bearing in mind the uncertainties attached to the measured parameters of coals.

## VI. BEHAVIOUR ON HEATING

An additional test to be applied to model structures is how such molecules would behave when pyrolysed. A very elegant study of the behaviour of a number of polymers has proved of great help in this respect. The polymers were made by condensation of various aromatic molecules with $^{14}$C-labelled formaldehyde which permitted the subsequent fate of the methylene linkages to be determined. The linkages in these polymers are not of the 9:10 dihydro type but the results of the work are still of interest and relevance.

In those cases where the aromatic molecules are unsubstituted hydro-carbons (naphthalene, anthracene, pyrene, coronene, etc.), the primary product of pyrolysis between 400 and 500°C is almost exclusively tar, as a consequence of random breakage of methylene bridges and dispro-portionation of the hydrogen. Recondensation of residues which are deprived of hydrogen occurs. Between 500 and 1000°C, when all the methylene hydrogen is effectively gone, aromatic hydrogen is evolved as further condensation occurs. Although some features of the process are similar to what happens in coal pyrolysis, the stages of tar and gas evolution are too sharply separated and the yield of tar is too high.

The presence of alkyl substituents on the aromatic molecules modifies the process by the evolution of some gas during the first stage of pyrolysis, but a more striking difference occurs when the aromatic molecule used carries hydroxyl substituents. These are detached by condensation reactions which compete with the depolymerization of methylene bridges and lead to a marked reduction in the tar yield. The resemblance to the behaviour of coal is thus increased if both hydroxyl and alkyl groups are present: tar, gas and water are evolved between 400 and 500°C and mainly hydrogen gas thereafter. (Quinone groups are likely to behave similarly to hydroxyl groups.)

It may therefore be concluded that the proposed types of model structure would behave in at least roughly the right manner during pyrolysis, as far as evolution of decomposition products is concerned. As described in Chapter 1, another important characteristic of coking coals is that they become fluid at temperatures between 400 and 500°C and swell up because of gas evolution, so forming a coherent coke. In the same study of the behaviour of aromatic polymers, measurements were made of the dilatation curves of polymers and it was found possible to simulate the dilatation behaviour of a coking coal with polymers made from phenanthrene and pyrene as in Fig. 18.

However, this was only possible over a limited range of degrees of polymerization: the number of bridges per monomer (b) must be only slightly greater than unity for the right form of dilatation curve to be obtained. (Solubility of the polymer in solvents also changes at similar b values from complete solubility when b is less than unity to partial solubility when b is slightly greater than 1 and to insolubility for higher values.)

The direct application of this finding to coal is unwise in view of the different bridging links postulated, and would in any case be feasible only if a distribution of molecular weights could be invoked. The parameter b would be approximately two for both the models developed if it is assumed that a 9:10 dihydro linkage is equivalent to two methylene bridges—but there is no proof of this. It is therefore not possible as yet to use such models to account for the coking characteristics of coals of different ranks. However, since it has been shown that dilatation behaviour is very sensitive to changes in the parameter b, a small reduction in the degree of cross-linking could account for the observed differences. The infra-red spectrum indicates a lower degree of substitution of the aromatic rings in coking coals which probably implies fewer cross-links. Other factors might also be involved: it has been shown that the degree of parallel orientation of layers is greater in a coking coal than in a low-rank coal and the porosity is lower. Both these changes could

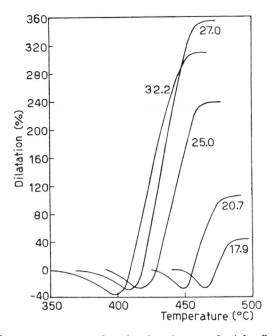

*Fig. 18a. Dilatometer curves of coals of various ranks (the figures indicate the percentage volatile matter).*

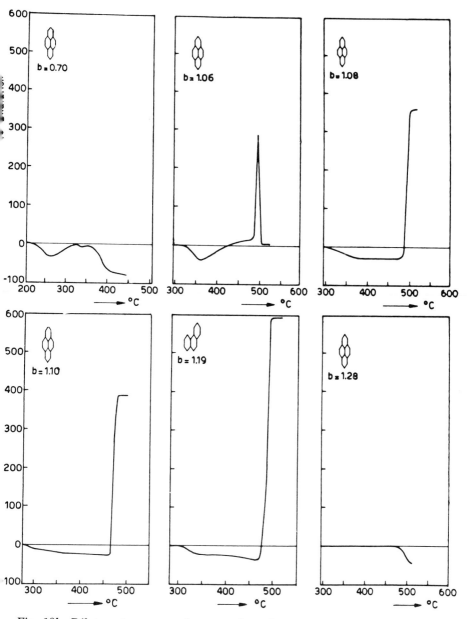

Fig. 18b. Dilatometer curves of some polycondensation products of pyrene and phenanthrene with formaldehyde (b = number of bridges per monomer unit).

(Figs 18a and b reproduced with permission from van Krevelen, 1961.)

impede the recombination of depolymerized material. Again, for appreciable fluidity the system must be plasticized by the lower molecular weight fragments of decomposition and their rate of evaporation must therefore not be too rapid.

The investigation of the structure of coal has been difficult, but at the same time challenging. Much effort has been spent in the exploration of individual features of the structure, and several attempts have been made to combine the results into model structures. There is a danger that the construction of models may lead to undue confidence that the chemical constitution of coal is understood; the main value of the exercise lies in the demonstration that there are many types of structure which do not occur to a marked degree in coal.

The work described in this chapter has formed the subject matter of a very large number of publications. Readers who wish for a fuller account of the subject with detailed references to the original papers should consult the references marked with an asterisk below. For details of more recent research, the reader should consult the journal *Fuel* and *Fuel and Energy Abstracts*.

## REFERENCES

Blom, L. (1960). Thesis, Delft.
Brown, J. K. (1955). *J. Chem. Soc.* 744-752.
Diamond, R. (1957). *Acta Crystallogr.* **10,** 359-364.
*Francis, W. (1961). "Coal: Its Formation and Composition." (2nd ed.) Edward Arnold (Publishers) Ltd., London.
Hirsch, P. B. (1954). *Proc. R. Soc.* A **226,** 143-169.
Hirsch, P. B. (1958). *In* "Residential Conf. on Science in the Use of Coal." pp. A29-33. Inst. of Fuel, London.
Juckes, L. M. and Pitt, G. J. (1977). *J. Microsc.* **109,** 13-21.
*Lowry, H. H. (Ed.) (1963). "Chemistry of Coal Utilisation." (Supplementary vol.) John Wiley & Sons, Inc., New York and London.
Retcofsky, H. L. and Friedel, R. A. (1970). *In* "Spectrometry of Fuels." (R. A. Friedel, Ed.) pp. 70-89. Plenum Press, New York and London.
Retcofsky, H. L. and Friedel, R. A. (1976). *Fuel, Lond.* **55,** 363-4.
Schuyer, J. and van Krevelen, D. W. (1954). *Fuel, Lond.* **33,** 176-183.
*van Krevelen, D. W. (1961). "Coal. Typology-chemistry-physics-constitution." Elsevier, Amsterdam.
van Krevelen, D. W. and Schuyer, J. (1957). "Coal Science." Elsevier, Amsterdam.

# 3 Carbonization and Coking

## J. GIBSON

*Board Member for Science, National Coal Board*

## I. INTRODUCTION

Carbonization is the process of destructive distillation of organic substances in the absence of air to yield a more or less pure carbon while producing liquid and gaseous products. Wood, sugar and vegetable matter yield charcoal, whereas coals of certain rank produce coke. This coke is used as fuel and as a metallurgical reducing agent mainly in the iron and steel industry where the greater proportion is employed in blast furnaces ($\sim$80%). Liquid by-products, after processing, yield a wide variety of chemicals; the residue and the gas are used as fuels.

In the seventeenth century, demands for wood for charcoal making began to outstrip supply and recourse was made to untreated coal to supplement the growing need for metallurgical fuel. The results were disappointing and further investigation led to the carbonization of coal by adapting charcoal production methods in which coal was carbonized in rounded heaps. Coke was widely produced in British coalfields in the seventeenth century and initially used for blacksmiths' fires and malting.

Abraham Darby of Coalbrookdale in Shropshire successfully used coke for iron smelting early in the eighteenth century. Darby's process was readily adopted and 81 coke blast furnaces were built in Britain in one year, 1790-91.

The primitive process of heating coal in rounded heaps—the hearth process—remained the principal one for over a century, although an improved form of oven in the shape of a beehive was developed in Newcastle

51

in about 1759. Beehive coke and charcoal are still used in the iron industry in many parts of the world. However, when indirectly heated slot ovens were introduced in the nineteenth century it became possible to collect and use the by-products. Development of the by-product recovery oven, which led to the modern coke oven, gave further impetus to the industrial revolution which had started at Coalbrookdale with Abraham Darby.

Next to combustion, carbonization is the largest use made of coal. Production has grown from a few thousand tonnes a year in Darby's day to a world total of 371 million tonnes in 1974. To a much lesser extent other feedstocks are carbonized. Petroleum refinery residues yield petroleum coke used to make arc steel electrodes and anodes for aluminium smelting. Vegetable matter from various sources and organic compounds are converted into active carbons for specific uses, and so on. This chapter will deal with coal carbonization and how it has been developed to the extent that a wider range of coals can now be used in coke making.

## II. CARBONIZATION PROCESSES

For details of coke manufacture reference should be made to books in the reading list. Table I summarizes the processes commonly used. It should be noted however that by far the greatest production is of hard coke for metallurgical purposes and with blast furnace applications predominating.

*Table I. Simple classification of carbonization processes according to temperatures to which coal is heated.*

| Carbonization process | Final temp. range °C | Aim and benefits | Examples |
|---|---|---|---|
| Low temperature | 450-700 | Reactive coke and high tar yield | "Rexco" (700°C) made in cylindrical vertical retorts. "Coalite" (650°C) made in vertical tubes. |
| Medium temperature | 750-900 | Reactive coke with high gas yield, or domestic briquettes. | Town gas and gas coke (obsolete) "Phurnacite", low volatile steam coal, pitch-bound briquettes carbonized at 800°C. |
| High temperature | 900-1050 | Hard, unreactive coke for metallurgical use. | Foundry coke (900°C) Blast furnace coke (950-1050°C). |

## High temperature carbonization

The equipment used for making foundry or blast furnace coke invariably consists of a battery of 10 to 100 slot ovens side by side indirectly heated through the oven walls by hot gases produced from burning the gases evolved during carbonization. Each oven may be charged with over 20 tons of coal and may be about 14 m between doors, up to 6·5 m high, average width 0·4-0·6 m with a tapering of 50-100 mm to facilitate discharge of the coke (typical dimensions in modern designs). Removable doors at each end seal the oven to prevent ingress of air and leakage of gas during the coking period. A cross-section of a typical oven is shown in Fig. 1.

The refractory linings of oven walls, flues and regenerator walls are made from silica bricks since they encounter the highest temperatures and the most severe mechanical stresses. The checker bricks in the regenerator are of fireclay. The battery is enclosed within a layer of insulating brick surrounded by common brick.

The coal to be carbonized is stored in overhead charging bunkers located at each end of the battery and a pre-determined weight is fed into a charging car which travels over the length of the battery to fill the individual ovens through charge holes in the oven top. The sequence of charging, levelling and replacing lids is completed as quickly as possible within a few minutes.

Coking is completed in about 12 to 80 hours depending upon oven width, flue temperatures used and the type of coke required. The coke is then pushed from the oven by a ram into a quenching car. A programme of charging, carbonizing and discharging of the ovens within a battery is arranged so that during its life the whole battery is maintained at a fairly even temperature and coke making is as near continuous as possible.

## Coke briquettes and formed coke

Coke in the form of briquettes is manufactured from coals normally considered to be too poorly endowed for use as a main component in coking blends. They are manufactured for use as a domestic fuel (e.g. 'Phurnacite') or as metallurgical fuel (e.g. F.M.C., Bergbau-Lurgi or Ancit processes), usually called "formed coke" in the latter case.

'Phurnacite' is an example of the, at present, more widely used process (Fig. 2). Non-caking coal of low-volatile content (CRC 201a—see Chapter 1, Fig. 8) is briquetted with a binder and carbonized in ovens at 750-800°C. Anthracite may be added and the process can be modified to process coals of higher volatile matter content or caking capacity after modification by oxidation or chemicals (e.g. iron oxide or lime) to prevent swelling and sticking together during carbonization. As a variation, low-volatile char prepared by pyrolysis of high-volatile coal may be briquetted with binder prepared from the pyrolysis products as in the FMC process. Treatment to

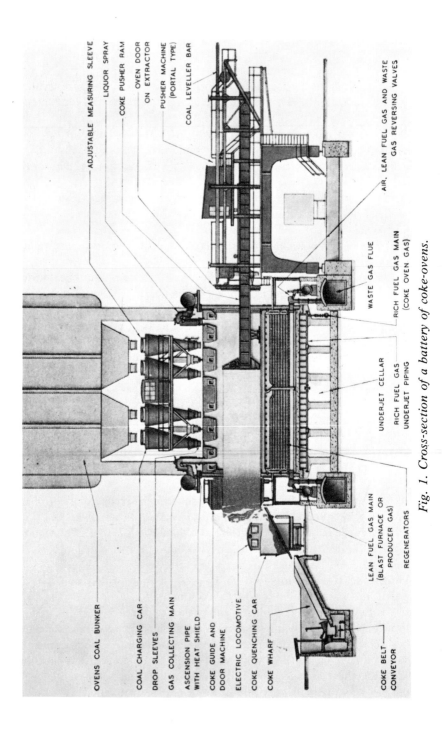

OVENS COAL BUNKER

COAL CHARGING CAR

DROP SLEEVES

GAS COLLECTING MAIN

ASCENSION PIPE
WITH HEAT SHIELD

COKE GUIDE AND
DOOR MACHINE

ELECTRIC LOCOMOTIVE

COKE QUENCHING CAR

COKE WHARF

COKE BELT-
CONVEYOR

ADJUSTABLE MEASURING SLEEVE

LIQUOR SPRAY

COKE PUSHER RAM

OVEN DOOR
ON EXTRACTOR

PUSHER MACHINE
(PORTAL TYPE)

COAL LEVELLER BAR

AIR, LEAN FUEL GAS AND WASTE
GAS REVERSING VALVES

WASTE GAS FLUE

RICH FUEL GAS MAIN
(COKE OVEN GAS)

UNDERJET CELLAR

RICH FUEL GAS
UNDERJET PIPING

LEAN FUEL GAS MAIN
(BLAST FURNACE OR
PRODUCER GAS)

REGENERATORS

*Fig. 1. Cross-section of a battery of coke-ovens.*

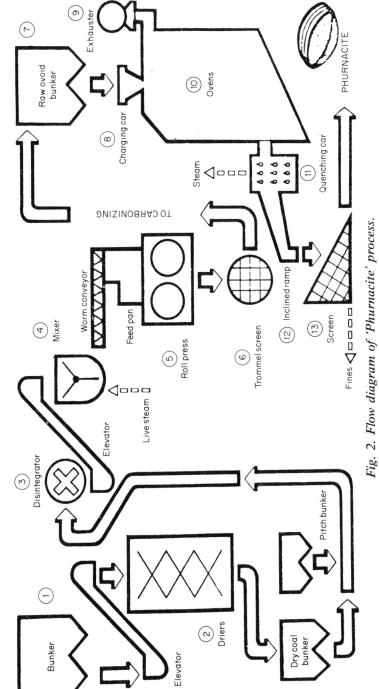

Fig. 2. Flow diagram of 'Phurnacite' process.

higher temperatures is necessary to produce blast furnace formed coke, otherwise the product can be used as a domestic fuel or metallurgical reductant where strength and volatile matter content are not so critical.

In an alternative method of manufacture, coal is partially carbonized at temperatures between 400 and 700°C depending upon its properties and then hot briquetted with caking coal as a binder. These briquettes have been used in the 'green' condition in blast furnace trials without adverse effects. Demonstration plants are in production in Germany and the British Steel Corporation are building one near Scunthorpe, which incorporates an extra heat treatment furnace for improving the mechanical properties of the product if it proves necessary.

## III. STRUCTURE OF COAL AND THE MECHANISM OF CARBONIZATION

### Thermal and rheological behaviour of coal

Upon heating, coal undergoes chemical changes giving rise to the evolution of gas and condensible vapours leaving behind a solid residue consisting almost entirely of carbon. In the temperature range 350-500°C, depending upon the rank of the coal, coking coals soften, become plastic, coalesce into a coherent mass which swells and then forms a solid porous structure. In this series of transformations from coal, two important temperature zones can be distinguished. The first is that in which the coal is plastic and the second is that at higher temperatures in which the resolidified material contracts.

### 1. Plastic temperature range

Prime quality coking coals have volatile contents in the range 20-32% (d.m.m.f.), become plastic before active decomposition occurs (Fig. 3) and are so named because they yield strong coke with good abrasion resistance. Viscosity and rate of devolatilization of the plastic mass are such as to minimize intragranular swelling but enable neighbouring coal particles to adhere strongly. The coke thus formed has fairly uniform pores of small diameter surrounded by relatively thick walls giving rise to a high resistance to abrasion.

Coal with volatile matter contents outside the range 20-32% undergo decomposition both before and during the plastic temperature zone. In many cases strong, intragranular swelling occurs and the whole mass may foam, giving rise to high porosity and thin walled pores of large diameter within the coke. This product has a relatively low resistance to abrasion. It is clearly seen that the inherent strength and abrasion resistance of coke are determined by the behaviour of the parent coal in the plastic zone.

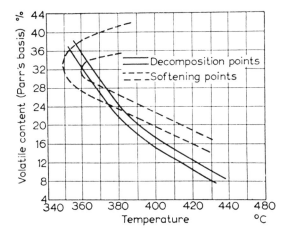

Fig. 3. Volatile matter of coal in relation to the softening and decomposition temperatures.

Figure 4 shows the contraction/expansion behaviour as determined in low temperature dilatometry (for expediency it is usual to employ different dilatometers in the plastic and contraction temperature zones).

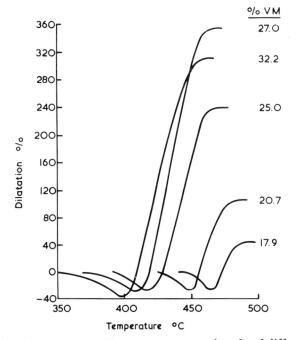

Fig. 4. Low temperature dilatometer curves of coals of different rank.

## 2. Post-plastic (contraction) temperature range

As the temperature is raised after the plastic mass resolidifies, the solid contracts but at a rate which is non-uniform and dependent upon the rate of devolatilization and reorientation processes in the solid, varying from coal to coal. If coefficients of contraction measured in a high temperature dilatometer are plotted against temperature two peaks are observed; the first occurs just after resolidification ($\sim$500°C) and the second in the region of 750°C (Fig. 5). (The coefficient of contraction of a pencil of coke of length l is defined as

$$\frac{1}{l_0} \times \frac{dl}{d\Theta}$$

where $l_0$    = length of sample at resolidification

and    $\dfrac{dl}{d\Theta}$ = slope of curve at temperature $\Theta$).

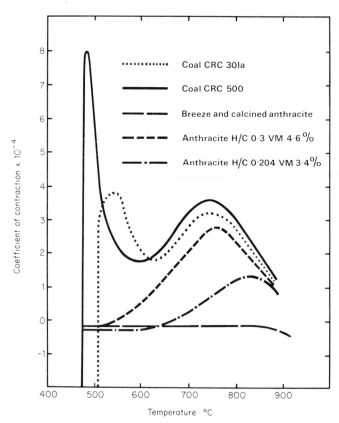

*Fig. 5. Contraction of coal in high temperature dilatometer.*

In a coke oven where heat is supplied from the oven walls the plastic layer travels through the coal charge from each wall leaving behind it an apparently solid but visco-elastic semi-coke undergoing further solidification and contraction. While the moving plastic layer is expanding to form the porous semi-coke structure, the already solidified semi-coke nearer the oven wall is undergoing further decomposition and contracting unevenly to form the denser, strong, higher temperature coke at the completion of carbonization usually in the range 900-1050°C. Thus, alternating compressive and tensile forces are produced in the semi-coke as shown schematically in Fig. 6. Differential strains are set up and result in the formation of fissures which break up the mass. A primary fissure network is associated with the rapid contraction at about 500°C and it has been shown that the size of coke discharged from the oven is related to the height of the first peak in the

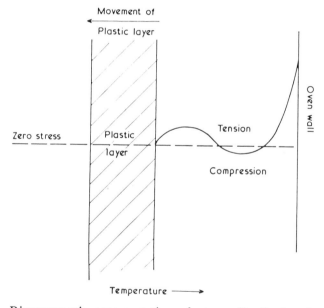

*Fig. 6. Diagrammatic representation of stress distribution in semicoke during carbonization (reproduced with permission from Patrick, 1974).*

contraction curve. A secondary fissure network is apparent if the coke pieces are sectioned. On impact these secondary fissures extend to break up the coke into smaller pieces. The impact strength of coke is seen to be related therefore to the height of the second peak in the contraction curve.

Having described the physical changes taking place, it is useful to consider broadly the accompanying chemical and structural aspects. The plastic stage involves the breaking of cross-linkages, comprising either oxygen or non-aromatic carbon bridges between neighbouring aromatic groups, leading to mobility of some of the decomposition products. The lower molecular weight components can undergo further change to yield (a) the gaseous volatile matter consisting of hydrogen, methane and other hydrocarbons, and (b) the highly complex mixtures found in coal tar. The higher molecular weight fractions remain and form semi-coke on solidification, during which adjacent aromatic clusters join up through a free radical condensation mechanism. These free radicals are formed from the initial scissions and side chains broken off through heat treatment.

The semi-coke has visco-elastic properties up to 700°C and transforms into hard coke, a brittle solid, as the temperature rises to about 1000°C. The process is accompanied by elimination of hydrogen and growth of graphite-like layers arising during the plastic stage and resolidification to semi-coke. The predominating gaseous product is hydrogen, removed from the periphery of the aromatic layers which can link together in the manner shown in Fig. 7. X-ray diffraction, magnetic susceptibility and electron microscope data confirm layer growth arranged in graphite type crystallites at these relatively low temperatures.

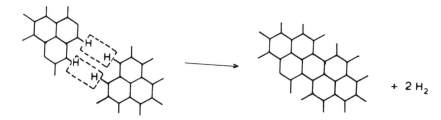

*Fig. 7. Hypothetical condensation process leading to layer growth (reproduced with permission from Patrick, 1974).*

## Petrographic aspects

Coal is far from being a homogeneous substance. It has been shown in Chapter 2 that examination of thin sections or polished blocks under the microscope reveals that it is a complex mixture of plant debris which has undergone transformation into coal in the course of geological time. A simplified classification of coal for coking technology is given in Table II.

The major part of the coals used for carbonization is made up of 'reactive' material, i.e. material which, on heating in the coke oven, reacts to the conditions in a definite pattern as the temperature rises; vitrinite and exinite are examples. These constituents of the coal undergo the softening stage in the 400-500°C range, and act as the binder for the 'inert' material. The 'inert' material does not soften, nor does it undergo subsequent contraction to the same extent as the 'reactive' components; it does, however, lose volatile matter and suffer internal structural chemical change. The role of the "inert" material in the coking process is to reduce the overall swelling and contraction. If good quality coke is to be produced, a balance in terms of content and size grading between the 'reactive' and 'inert' composition of the charge to the oven must be achieved.

*Table II. Petrographic classification of coal for coking technology.*

| Coking characteristic | Petrographic constituents |
|---|---|
| Reactive | Vitrinite with reflectance between 0·5 and 2·0%. Exinite |
| Intermediate (partially reactive) | Semifusinite (reflectance less than 2·0%). |
| Inert | Fusinite, micrinite. Vitrinite with reflectance greater than 2·0%. Mineral matter (shale, pyrite, etc.). |

## Kinetics of carbonization

The kinetics of coal carbonization have been studied by thermogravimetric, thermovolumetric and differential thermal analysis techniques. With the assistance of plasticity and dilatometric measurements, the reactions leading to the transformation of prime coking coal into coke have been postulated as follows:

(1) Coking coal $\xrightarrow{K_1}$ metaplast

(2) Metaplast $\xrightarrow{K_2}$ semi-coke + primary volatiles

(3) Semi-coke $\xrightarrow{K_3}$ coke + secondary gas

where $K_1$, $K_2$, and $K_3$ represent the relevant velocity constants.

Reaction 1 is a depolymerization reaction in which coal is transformed into an unstable intermediate responsible for the plastic behaviour. Reaction 2 represents the transformation of metaplast into semi-coke (resolidification) which is accompanied by gas evolution causing the mass to swell. In Reaction 3, the semi-coke structural units lose methane and hydrogen, and weld together to give a more compact and coherent coke.

This kinetic interpretation cannot provide a complete description of the complicated coking process but can be regarded as a mathematical model, which lends itself to a semi-quantitative description of the experimental results and gives a satisfactory explanation of the physical phenomena occurring during the coking of coal.

## Carbonization of model substances

The theories briefly described above are difficult to test because coal is a complex and variable substance. Large, organic molecules of known constitution have therefore been used as models with which to study the mechanism of carbonization. The first used by Riley and his co-workers were polynuclear, aromatic dye-stuffs of the dibenzanthrone type, because in most respects they behave like bituminous coals upon carbonization. In more elaborate studies later workers used organic molecules containing hexagon nuclei of known structure linked with bridge structures (e.g. $CH_2$ groups) into which could be introduced radio-actively labelled carbon atoms to facilitate identification and quantitative assessment of the bridging elements during pyrolysis.

The results of model studies confirm the main features of the theories developed to explain the mechanism of carbonization. They can be best described in the words of van Krevelen (1961):

The basic process in carbonization is thermal cracking ('depolymerization') with simultaneous disproportionation (dismutation) of hydrogen. Fragments enriched in hydrogen evaporate as tar, the others recondense to yield semi-coke. Side chains are cracked at the same time. Aromatic carbon-hydrogen bonds are broken at higher temperatures (secondary carbonization). Functional groups, such as hydroxyls, give rise to a competing reaction, intermolecular condensation, which has a severe influence on the whole process, especially on the plastic properties during carbonization.

## Factors affecting fluidity and swelling

The normal behaviour of coals undergoing carbonization may be affected by oxidation or hydrogenation of the coal, as well as by the presence of mineral matter (particularly sulphur) in the coal. Higher heating rates increase fluidity and swelling, and the temperature range over which the coal is plastic is extended. Some coals, which show no softening when heated slowly, soften and may swell when heated more rapidly. Some authors have concluded that very fine grinding of coal reduces its plasticity, the effect being particularly pronounced with coals high in inert components. Crushing to the coarse sizes used in commercial plants has different effects with different coals, depending on their petrographic compositions.

Coals which soften, swell and form coke on carbonization under atmospheric pressure do not give a coherent coke when heated *in vacuo*. Reduced pressure decreases the degree of softening and swelling. On the other hand, if coal is heated under pressure the softening and swelling increase, and a firmer coke is produced. To obtain an appreciable effect, however, pressures of several atmospheres are necessary.

The plastic properties of coals are affected markedly by even mild oxidation. In general, oxidation restricts the plastic temperature range and brings about a decrease in fluidity and swelling.

Storage of coals, particularly those of higher volatile matter content, may result in oxidation, even at ambient temperatures. Oxidation of the coal is harmful for normal coking practice but, as noted previously, under controlled conditions it is useful in reducing the swelling of briquettes when these are carbonized.

Mild hydrogenation of coal increases the swelling and the fluidity; the plastic range is also widened. The reaction of sulphur with coal is similar to that of oxygen.

## IV. COALS AND BLENDS FOR CARBONIZATION

### Dependence on class of coal

In the United Kingdom, coals are classified according to the type of coke they produce when heated under specified conditions. The classification of the National Coal Board (see Chapter 1, Fig. 8) relates the rank of the coal with its caking capacity, as measured by the Gray-King test, and its volatile matter content on the dry, mineral-matter-free basis. Both tests are described by British Standard Specifications. There is an international classification system but unfortunately it is not so practicable and has failed to find favour.

In general, coals vary widely in their ability to make coke, and it is customary to blend them before carbonization in order to produce the type of coke required. Foundry coke, the hardest and most dense product required, has traditionally been made from prime coking coals, generally of CRC 301, of volatile matter contents in the range 19·6 to 32% (d.m.m.f.). Blast furnace coke, also required as a hard and strong fuel, is also made from these coals, but, because it is used in smaller sizes, higher volatile coals (>32% V.M.) of CRC 400 and 500 have been found suitable alone or in blends with prime coking coals. For the manufacture of town gas and the softer, reactive cokes, coals of lower rank have been extensively used. In particular, low-temperature carbonization processes for the production of reactive domestic fuel have favoured the use of the less strongly coking coals of CRC 600 and even CRC 700.

## Coking modifiers

The exhaustion of many of the prime coking seams all over the world has led, in the case of foundry and blast furnace coke making, to the use of coals formerly considered suitable only for gas making. The recent trend in the UK is shown in Fig. 8. The cokes made from these replacement coals (CRC 400-700) may be substantially improved by the incorporation of coking modifiers. It has been mentioned how dilatometric studies in the post-plastic

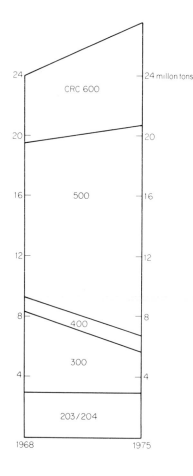

*Fig. 8. Trend in usage of coals for carbonization in UK.*

zone reveal the presence of two contraction rate peaks related to primary and secondary fissure-producing influences. The principal fissure-forming influence tends to control the size of pieces on discharge from the coke oven. The second affects the less severe system of fissures which only become manifest when subjecting the pieces so formed to more severe stresses, as in the shatter test; hence the relationships between the heights of the first and second peaks in the contraction-rate curve and the size of coke and shatter strength respectively.

It will be noted from Fig. 5 that neither coke breeze nor anthracite exhibits any contraction in the region of the first contraction-rate peak, while at or around the temperature of the second peak, anthracite does contract. If the above relationships are valid, then the addition of anthracite or breeze to a coking coal should reduce the first peak and increase the mean size of coke produced from such a mixture. Similarly, reduction of the second peak by addition of breeze should bring about improvement in the shatter index of the coke. Anthracite, however, which cannot affect the second peak to the same degree, should have a less noticeable effect. All these postulates have been verified experimentally. Furthermore, it was shown that calcining anthracite and reducing its volatile matter content progressively reduced its second contraction rate peak (Fig. 5). Comparison of cokes made without any addition, with untreated anthracite and with calcined anthracite, showed that untreated anthracite influenced only the mean size, whereas calcined anthracite increased mean size to a greater extent and improved impact resistance, thus confirming the suggested relationship. However, the amounts of breeze and anthracite which can be incorporated in a blend may be limited by their effect on abrasion resistance; both cause deterioration after certain levels of addition, depending upon their size grading. With high-volatile base coals the more fluid, low-volatile steam coals (CRC 204) can help to compensate for this, and where control of size, strength and abrasion resistance are necessary, these steam coals have an important function.

The size grading of the coking modifier is important and usually it is finely ground. Large irregular inert particles build up stresses and propagate cracks as the semi-coke contracts around them weakening the coke product and reducing its resistance to abrasion, thereby worsening rather than improving the properties.

## Preparation of the coal blend for carbonization

Blending of coals with appropriately different characteristics is practised to improve the physical and chemical properties of the coke product, to enable a proportion of inferior or marginal coking coals to be used and to control and improve the yield of coke. It also has other objectives, e.g. to prevent damage to the ovens during the carbonization of coals which develop a comparatively high pressure during coking, and to obtain maximum economic oven efficiency.

To ensure that a consistent product is obtained, careful segregation, strict proportioning and effective mixing of the component coals are essential. The cleaned coals are usually passed through 12·5 or 25 mm sieves before arrival at the coke ovens. In general the coals are processed as they arrive at the coke ovens; some stock may be maintained, but the coking properties of coal deteriorate on storage and unless adequate precautions can be taken to prevent oxidation, the quantity stored is usually small. If a stock is maintained, it must be consolidated, limited in height and covered with an impervious coating of tar or a suitable emulsion to prevent the entry of air.

Normally, different coals are bunkered separately. From the bunkers, the required proportion of each coal is measured out, using rotating tables, weigh belt or fixed-volume-delivery proportioners before passing to a crusher. Generally, the coals are crushed in a hammer mill to give a product containing 80-90% below 3 mm in size. The hammer mill acts also as a mixer, and this may be the only mixing required if the blend consists solely of coking coals. When contraction-modifiers such as coke breeze, anthracite or other low-volatile coals are incorporated, finer grinding and more efficient mixing are necessary, and mechanical mixers are recommended. A flow diagram of the coal preparation section of a modern plant producing foundry coke is shown in Fig. 9. In the example chosen, a large proportion of high-volatile coal (CRC 500) was used to replace prime coking coal no longer available. Coking modifiers—low-volatile coking coal (CRC 204) and finely ground coke breeze—enable high quality foundry coke to be produced equal to that made previously.

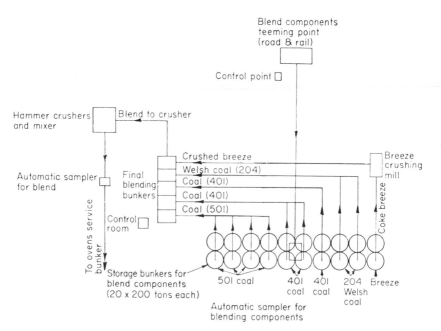

*Fig. 9. Flow diagram of a coal blending plant.*

The degree of crushing of the blend is often relaxed in the production of domestic cokes and town gas, so that only 60-70% may be less than 3 mm in size. In some low temperature processes, graded coals, e.g. 38 mm × 50 mm or 38 mm × 100 mm are processed without crushing.

## Further treatment

Before carbonization the blend may undergo further treatment to enhance its ability to make better coke or to improve the economics of manufacture.

As normally prepared, blends may contain as much as 10% free moisture by weight which impedes the flow properties and lowers bulk density. The throughput of the coke oven depends on the amount of coal charged, and on the duration and frequency of the carbonization cycle. Flow properties and the bulk density of the charge may be improved simply by spraying with a light oil at a rate of about 3 litres per tonne of coal.

Significant additional benefits accrue from heating the charge before it enters the oven; bulk density is increased and the carbonization time is reduced by removal of the water. Gains up to 50% in oven throughput have been reported when the charge is preheated to temperatures of 200°C. The abrasion resistance of the coke is usually improved by this treatment. Preheating is carried out in dilute-phase (entrainment) or dense-phase (fluidized bed) equipment, but it is essential to avoid oxidation of the coals to prevent loss of coking properties. Some coals make better coke if the charge is precompacted (stamp-charging). This method is not popular because it involves complicated additional equipment, although recent work is reported on combining preheating with stamp-charging.

## V. FURTHER READING

### Chemistry and physics of carbonization

van Krevelen, D. W. (1961). "Coal. Typology-chemistry-physics-constitution." Elsevier, Amsterdam.
Lowry, H. H. (Ed.) (1963). "Chemistry of Coal Utilisation." (Supplementary vol.) Wiley, New York.
Patrick, J. W. (1974). *Sci. Prog., Oxford* **61**, 375-399.

### Carbonization technology

Lowry, H. H. (Ed.) (1945) "Chemistry of Coal Utilisation." (2 vols, supplementary vol., 1963) Wiley, New York.
Wilson, P. J. and Wells, J. H. (1950). "Coal, Coke and Coal Chemicals." McGraw-Hall, New York.
McNeil, D. (1966). "Coal Carbonization Products." Pergamon Press Ltd., London.

# 4 Structure and Properties of Coal Derivatives

## G. J. PITT

*Coal Research Establishment, National Coal Board*

## I. INTRODUCTION

Many accounts have been written in the past about the very extensive range of derivatives obtainable from coal, including chemical compounds which are of great value; the distillation and refining of coal tar are of considerable commercial importance as will be seen in Chapter 9. It is, however, not unreasonable to look upon metallurgical coke as the principal coal derivative of commercial importance at the present time, since hundreds of millions of tonnes of coal are converted to coke per year, almost entirely for use in the manufacture of iron and steel. Coke is a chemical reductant and should be of suitable reactivity, but, in addition, it must satisfy physical requirements such as lump size and strength to ensure optimum performance of the blast furnace or foundry. Because of the importance attached to these factors, it is appropriate to study the structure of coke and assess the influence of structural features on the properties. This is of special relevance when prime coking coals have to be replaced by blends of other coals. In this chapter the structural changes which occur when coal is converted to coke and graphite will be described, followed by a discussion of various aspects of the materials science of coke and, to conclude, similar ideas will be developed for coal-based graphite electrodes for use in the arc steel process.

## II. DEVELOPMENT OF STRUCTURE ON HEAT-TREATMENT OF COAL

During carbonization, a coking coal softens and swells as gas and tar vapour are evolved within the particles. Figure 1 shows a cross-section of a bed of coal which has been heated from one side and allowed to cool, leaving a 'frozen' record of the state of the material after reaching the temperatures shown. Adjoining particles fuse together and a continuous porous material is

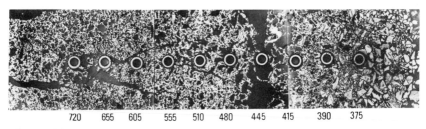

720    655    605    555    510    480    445    415    390    375

*Fig. 1. Photomicrograph of transformation of coal to coke in 'micro-coke-oven' (figures indicate maximum temperature reached (°C)).*

formed on resolidification. This semi-coke undergoes contraction and fissuring as further volatile products are evolved, leaving a coke which typically might contain 96%C and 0·5%H (the residue from the mineral matter in the coal being excluded).

X-ray analysis of the type described in Chapter 2 (Diamond, 1970) gives evidence that the degree of parallel stacking of the aromatic layers increases up to about 500°C (Fig. 2) coupled with a decrease in the proportion of 'amorphous' material (Fig. 3), substantiating the suggestion that in the plastic stage breakage of some cross-links and the evolution of lower

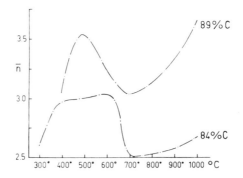

*Fig. 2. Mean number of layers per stack in a carbonized caking coal (89%C) and coking coal (84%C) (reproduced with permission from Diamond, 1960).*

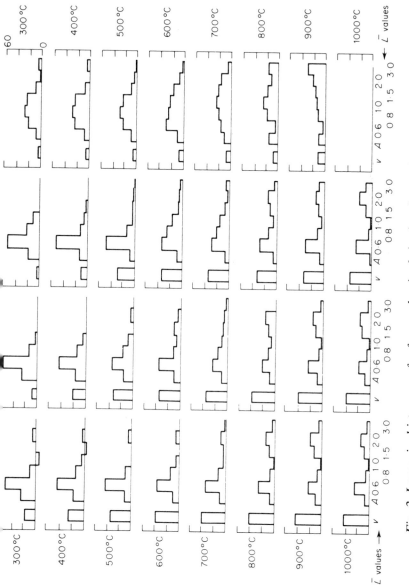
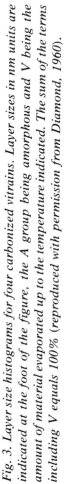

Fig. 3. Layer size histograms for four carbonized vitrains. Layer sizes in nm units are indicated at the foot of the figure, the A group being amorphous and V being the amount of material evaporated up to the temperature indicated. The sum of the terms including V equals 100% (reproduced with permission from Diamond, 1960).

molecular weight material remove some of the obstacles to stacking. At this stage there is no indication of growth of the layers (Fig. 4).

Between 500 and 700°C the layer diameter does start to increase, and this development is accompanied by a decrease in the stack height. Over this interval of temperature hydrogen is evolved and the layers begin to join up, but the decrease in the stack height indicates that layer growth is only possible at the expense of the parallelism. At temperatures above 700°C the layer diameter continues to increase and the stack height increases again. It might be said that the sort of order present in the semi-coke at 500°C is a false start and that progress to a higher degree of order is only possible by partial reorganization of the parallel stacks.

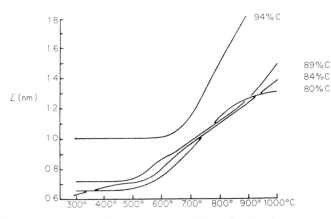

*Fig. 4. Variation of mean layer diameter with carbonization temperature for the four vitrains in Fig. 3 (reproduced with permission from Diamond, 1960).*

Although a higher degree of order exists in a coke prepared at about 1000°C, it is important to realise that the structure is still far from ordered in a crystalline sense (Fig. 5). The stacks of aromatic layers are still fairly small and moreover they are not mutually well-orientated or positioned for further growth, though cokes from medium-rank coals have a better medium-range order than those from low-rank coals or anthracites (perhaps because of their passage through a plastic state). The importance of this difference becomes apparent when the cokes are heated to even higher temperatures. Both the layer diameter and the stack height increase much more rapidly in the medium-rank coke than in the low-rank coke, so much so that cokes (and carbons in general) can be classified either as graphitizing or non-graphitizing (Fig. 6). The distinction arises partly from the degree of medium-range order in the coke, because reorientation of grossly misaligned stacks would be virtually impossible, and partly from the degree of (three-dimensional) cross-linking. Anthracite cokes are an intermediate case with reasonably good medium-range order, but a fairly high degree of cross-linking; they behave as if non-graphitizing at lower temperatures but finally graphitize at high temperatures when the cross-links are broken.

A new technique likely to be useful in supplementing X-ray analysis in the study of the 'molecular' structure of cokes and similar materials is high-resolution electron microscopy, which forms the subject of Chapter 5.

The information about the structure of coke which has been derived from X-ray analysis needs to be supplemented because cokes are observed to have

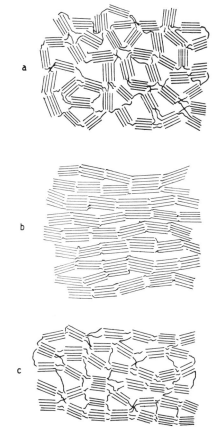

*Fig. 5. Schematic representation of the structures of cokes made from* (a) *a low-rank coal,* (b) *a coking coal and* (c) *an anthracite* (*reproduced with permission from van Krevelen, 1961*).

a texture when viewed under an optical microscope (Patrick *et al.*, 1973). A polished section through a coke made from a medium-rank coal, when examined between crossed polarizers, presents an appearance (Plate 2) which is described as a mosaic texture, with each unit area having an extinction direction different from those of its neighbours.

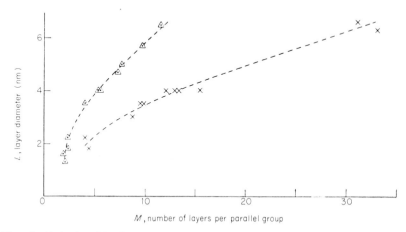

*Fig. 6. Relationship between the mean layer diameter, L, and the mean number, M, of layers per stack for graphitizing ($\times$) and non-graphitizing ($\triangle$) carbons (reproduced with permission from Franklin, 1951).*

The texture depends on the rank of the coal. A prime coking coal gives a mosaic size usually of the order of 1 $\mu$m, and with decreasing rank, caking coals have progressively smaller sized mosaics (Fig. 7). The coke made from low-rank coal appears isotropic, but it is not known whether it is truly isotropic or has a mosaic size too small to be resolvable by an optical microscope. In a coke made from a high-rank coal the mosaic effect is replaced by a lamellar texture (Plate 3), and in anthracite cokes there is overall anisotropy without any marked texture.

When coal is heated in a hot stage microscope the texture can first be observed at approximately the resolidification temperature and it does not change in pattern, though it becomes more intense, at higher temperatures. By analogy with what is observed when pitch is heated, it has been suggested that towards the end of the plastic stage the decomposing coal undergoes a transition to a state similar to a nematic liquid crystal, the so-called mesophase (White, 1975), which develops from a number of centres and spreads (Plates 4 and 5) until it coalesces throughout the mass, giving rise to the mosaic texture. While it might seem reasonable to postulate that coal behaves similarly to pitch, the mesophase is not normally observed when coal

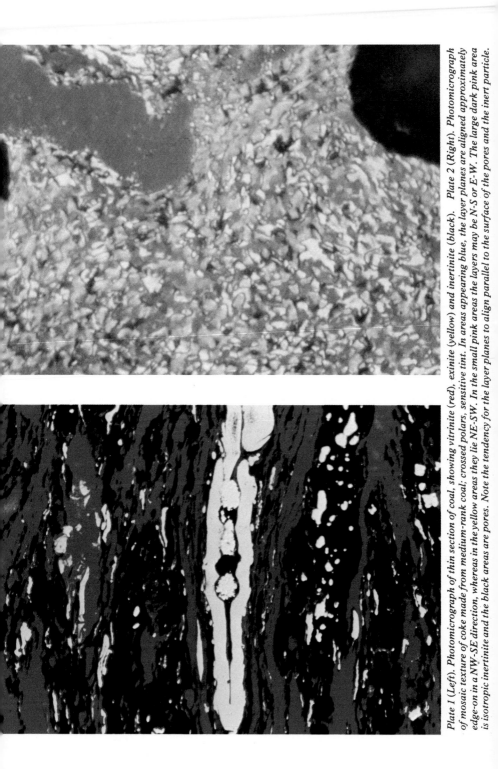

Plate 1 (Left). Photomicrograph of thin section of coal, showing vitrinite (red), exinite (yellow) and inertinite (black). Plate 2 (Right). Photomicrograph of mosaic texture of coke made from medium-rank coal; crossed polars, sensitive tint. In areas appearing blue, the layer planes are aligned approximately edge-on in a NW-SE direction, whereas in the yellow areas they lie NE-SW. In the small pink areas the layers may be N-S or E-W. The large dark pink area is isotropic inertinite and the black areas are pores. Note the tendency for the layer planes to align parallel to the surface of the pores and the inert particle.

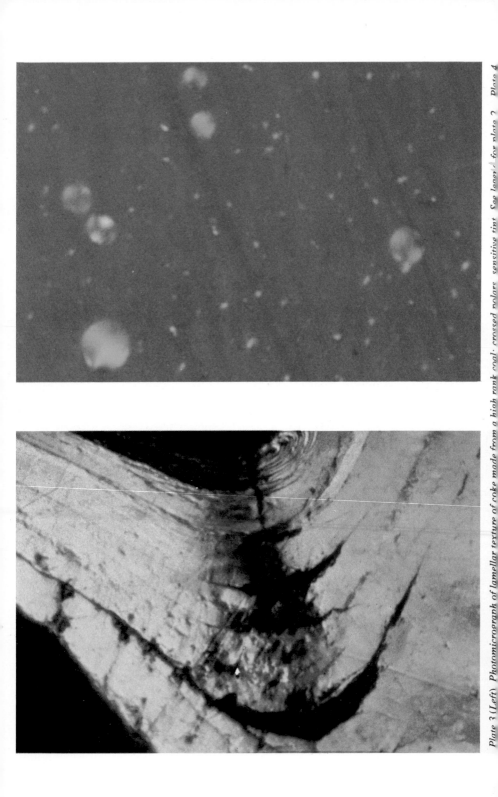

Plate 3 (Left) Photomicrograph of lamellar texture of coke made from a high rank coal; crossed polars, sensitive tint. See legend for plate 2. Plate 4

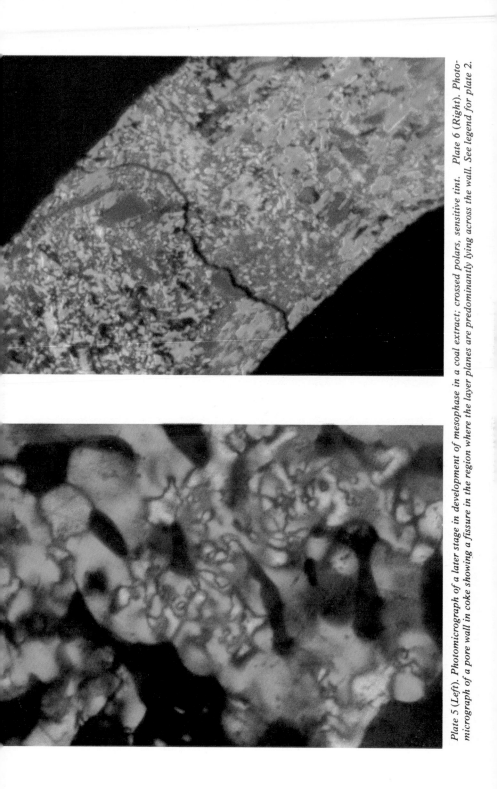

Plate 5 (Left). Photomicrograph of a later stage in development of mesophase in a coal extract; crossed polars, sensitive tint.    Plate 6 (Right). Photomicrograph of a pore wall in coke showing a fissure in the region where the layer planes are predominantly lying across the wall. See legend for plate 2.

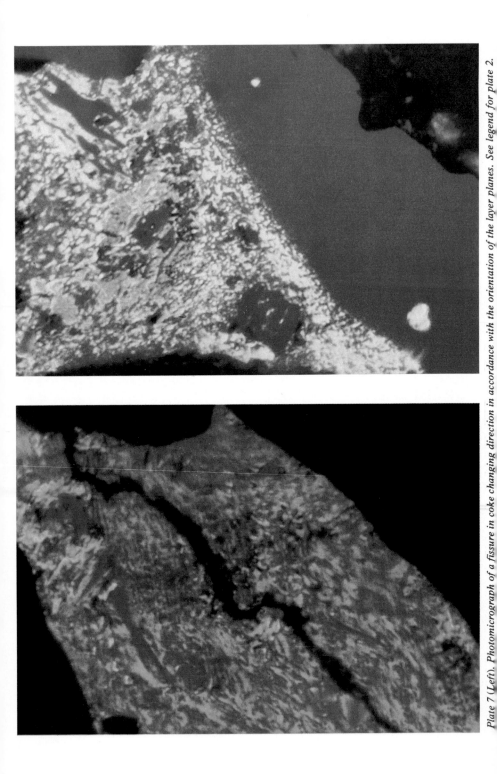

*Plate 7 (Left). Photomicrograph of a fissure in coke changing direction in accordance with the orientation of the layer planes. See legend for plate 2.*

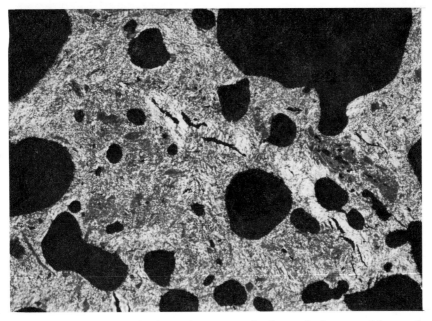

*Fig. 7. Photomicrograph of mosaic texture of coke made from a caking coal; crossed polars.*

is heated and so its occurrence during the formation of the anisotropic texture of coke is still hypothetical. Whatever the mode of formation of the texture, it represents a degree of preferential orientation of the stacks of aromatic layers extending over some hundreds of stacks in a given direction, since a mosaic unit of 0·5 μm (500 nm) diameter is made up of stacks of layers of diameter about 1 nm as determined by X-ray analysis.

## III. RELATIONSHIPS BETWEEN STRUCTURE AND PROPERTIES OF COKE

### Macrostructure

The first property of importance to the steel-maker is the size of the coke lumps. The main fissure pattern which governs the coke size is clearly defined in the coke mass as it emerges from the oven (Figs 8 and 9) so the fissures must have arisen between the time of resolidification and the end of the heating period. As heat is transmitted from the outer walls of the oven to the centre of the charge, any given vertical slice of the charge is hotter on the side nearer the oven wall than on the side nearer the oven centre. Con-

Fig. 8. Coke discharged from a coke oven.

Fig. 9. Cross-section of a lump of coke showing rounded 'cauliflower' end (oven wall side) resulting from differential stresses, and primary fissures.

sequently, the outer side of the slice has contracted more (because of a greater extent of evolution of volatile products and a higher degree of ordering of the residue) than the inner side and is therefore in a state of tension. The stress is eventually relieved by fissuring and it can be shown mathematically that the mean spacing of the fissure mesh (and hence the coke size) can be calculated from parameters such as the tensile strength and creep of the coke, the contraction characteristics and the temperature distribution. This calculation was originally performed using rough estimates of these parameters with some measure of success in accounting for the coke size, and is now being refined using measurements of the tensile strength and creep made on the hot coke while it is being prepared.

When the coke leaves the oven it is quenched and transported to the top of the blast furnace where it starts to descend mixed with the iron ore. To permit efficient operation of the furnace there must not be excessive resistance to gas flow through the stack and this requires that there must not be an excessive proportion of fine particles in the coke. The steel-maker, therefore, wants a coke which will resist breakage and abrasion during transport and in the blast furnace.

Microscopic examination of a polished section through a lump of coke (Fig. 10) reveals a number of features of the structure which could influence

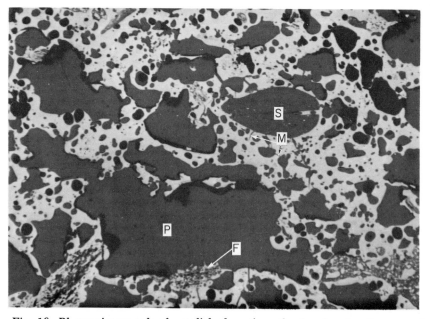

*Fig. 10. Photomicrograph of a polished section of coke showing features of the structure: P, pore; M, microfissure; F, fusain (carbonaceous inert material); S, shale (inorganic inert material).*

the strength, including pores, microfissures, inert particles and the anisotropic texture. The structure is a highly complex one and very variable from
point to point even over small distances, in addition to showing an overall
variation from the outside to the centre of the oven associated with a marked
difference in the temperature history. These features will be discussed in turn
with a view to assessing their influence on the strength.

## Pores

If it is accepted that tensile strength is an adequate measure of the strength
of coke (and this can be contested), then the effect of pores can be
investigated experimentally. Discs of coke can be prepared of, say , 10 mm
diameter and their porosity determined by measurement of lump density and
true density. The tensile strength can then be measured by a diametrical
compression test. Large numbers of tests are necessary because of the
variability, but it can be shown that tensile strength tends to decrease as
porosity increases. This suggests that the strength of a piece of coke may be
proportional to the cross-section area of solid material available to bear the
load. In an idealized structure with pores of equal size distributed in a
regular cubic array (Fig. 11), the fracture will occur in one of the (100) planes
where the proportion of solid material is a minimum, giving $S/S_0 = 1 - kp^{\frac{2}{3}}$,
where $S$ is the tensile strength when the fractional volume porosity is $p$, and

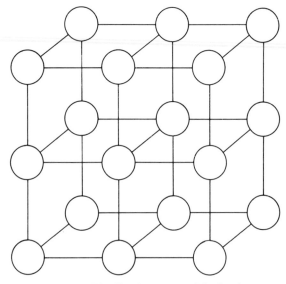

*Fig. 11. Idealized pore model of coke.*

$S_0$ is the tensile strength of pore-free coke. The numerical factor $k$ is $1\cdot21$ for the cubic array and would be $1\cdot11$ for a rhombohedral array; the formula quoted must be modified for fractional porosities greater than $0\cdot5$ (i.e. 50% volume porosity) because the pores would then overlap in the simple model. This is a relationship which can be tested by plotting $S$ against $p^{\frac{1}{3}}$: a straight line should be obtained with slope/intercept $= -1\cdot2$ approximately. This is found to be the case for a number of cokes, as shown by the example in Fig. 12, and $S_0$ can be obtained by extrapolation, leading to the

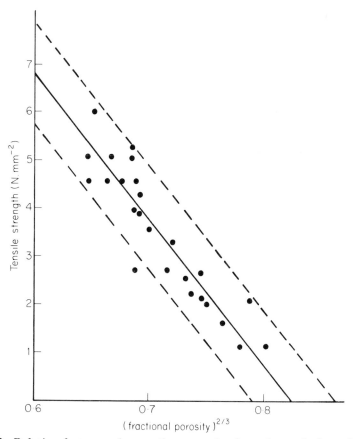

*Fig. 12. Relation between the tensile strength of a coke and the volume porosity.*

conclusion that porosity reduces the strength of typical cokes by a factor of between five and ten. There seems no likelihood that pore-free coke could be made economically, and if it were made it would probably not be of suitable reactivity for blast furnace use. However, the influence of porosity is so marked that even a small reduction could make a significant improvement in strength. In practice, the porosity can be reduced by increasing the bulk density of the coal charge.

It may be considered that the model pore structure chosen (Fig. 11) is over-idealized, and that it would be more realistic to assume a completely random distribution of pores without restriction on size or shape. This has also been tried, but the experimental data available do not permit a decision on which is the better model, although the regular array seems slightly preferable. The random pore model gives extrapolated strength values ($S_0$) about two thirds of those derived from the regular pore model, so the weakening effect of random pores is still sizeable.

At present it is not possible to state the effect of different sizes or shapes of pore, but the application of image-analysing computers now permits the collection of pore size distributions, shape factors and wall thickness distributions with sufficient accuracy to establish whether these characteristics also influence the strength.

## Microfissures and inert additives

Microfissures of various types occur in coke and it is often difficult to assess their effect. One particular type is caused by the inclusion of finely ground coke (coke breeze) in a blend of coals for coke making and this will serve as an example. During carbonization a breeze particle normally becomes engulfed by the fluid coal which then solidifies and contracts around it, generating stresses at the interface because the breeze has previously been fully carbonized and therefore does not contract. When the stress exceeds the yield stress of the semi-coke, a microfissure is formed (Fig. 13) and the stress level decreases.

By constructing a mathematical model in which spherical shells of semi-coke surround spherical breeze particles arranged in a regular array, it is possible to derive a theoretical relationship between the length of the microfissure and the size and concentration of breeze particles. Each particle is predicted to generate one microfissure, and the length of the microfissure should increase with the size of the breeze particle. Weight for weight, therefore, small particles should generate many more fissures (of shorter length) than large particles and this appears to be so in practice. Numerous short fissures could more effectively reduce the level of residual stress throughout the lump of coke and so increase its strength, whereas long fissures are potentially more weakening. This is in agreement with the observation that the breeze must be finely ground to be fully effective.

In practice, however, while the addition of breeze may increase the strength of the coke, it does not significantly increase its ability to withstand abrasion.

This may be explained by the characteristic way in which a microfissure creates a gap between a breeze particle and the matrix (Fig. 13), thereby increasing the amount of material which can be easily abraded from the surface over and above the local weakness resulting from the microfissures themselves. The addition of breeze to a coal blend alters the course of carbonization in a number of ways which might affect the strength of the final coke. It could be argued that the fissures are crack initiators (and therefore a source of weakness) which are an unwanted side-effect associated

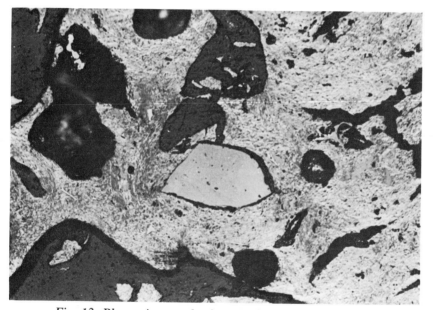

*Fig. 13. Photomicrograph of a coke breeze particle in coke.*

with some other wanted property of breeze such as increase in viscosity of the fluid coal. On the other hand, the fissures could act in a beneficial manner as stress relievers or as crack stoppers. Some evidence on this point may be obtainable by carrying out controlled fracture tests and observing the course followed by the fracture.

In contrast to coke breeze, other 'inert' constituents such as inertinite and shale have different effects. Inertinite undergoes some contraction during carbonization and is usually accommodated without fissuring, whereas shale generates wide fissures but forms a stronger bond to the coke and so bridges the fissures.

## Anisotropic texture

The last structural feature of coke to consider is the microscopic texture and the underlying anisotropy. It will be seen later that the properties of graphite parallel and perpendicular to the layer planes are very different, and, to a lesser extent, a similar directional difference must result from the anisotropic units responsible for coke texture. (Optical anisotropy is itself, of course, one such property.) In particular, the bonding is much stronger parallel to the layers than in the perpendicular direction. In a random mosaic, strong and weak regions will alternate along any line and the material will be equally strong in all directions. Sometimes, however, the mosaic texture develops a preferential orientation (while remaining essentially a mosaic) and then the coke will be weak in one direction (Plate 6). Examples of preferential orientation frequently occur adjacent to boundaries with pores (e.g. Plate 2) and in these cases the coke is better able to withstand tensile forces parallel to the boundary than perpendicular to it. Since fracture tends to be initiated by a surface flaw, the stronger bonding parallel to the surface hinders the propagation of a fissure.

The extent to which preferential orientation can influence fissure formation can be gauged from the example in Plate 7 which shows a fissure changing direction abruptly in response to a change in orientation. This means to say that more energy is used in creating fracture surface than would be the case if the fracture followed the shortest path, and the material becomes effectively stronger. In other materials it has been found that the extent to which a fracture 'follows' the structure depends on the rate of propagation. In impact breakage the crack travels at high speed and tends to ignore structure, whereas in controlled fracture the speed may be low enough for the crack to divert along the direction of least resistance. Coke texture may therefore be of greater importance in resisting, for example, thermal stresses rather than impact damage.

It may be of interest to mention that coke texture has provided some useful information on the carbonization of coal blends. If a mixture of two coals is carbonized, the temperature ranges over which they are plastic may or may not overlap. If they do not overlap, then each coal must act like an inert material to the other, but when the plastic ranges do overlap is there something akin to mutual solution of the two coals? The answer is found by examination of the coke texture (Plate 8). Areas are present which show the textures characteristic of the two individual coals and the boundaries between different textures are comparatively sharp; therefore very little interdiffusion occurs. (Gryaznov and Kopeliovich, 1975, have shown theoretically from the Brownian movement and the time during which the coals are plastic, that interdiffusion could occur to a depth of no more than about 2 $\mu$m in a typical case.) It can therefore be concluded that coke made from a blend of coals is held together by adhesion rather than homogenization.

# IV. THE STRUCTURE AND PROPERTIES OF GRAPHITE ELECTRODES

The prospective use of solvent extracts of coal to make graphite electrodes for arc steel furnaces is described in Chapter 9. Without going into the details of the various stages of the process, it is sufficient to note that the extract is converted to coke which is ground, mixed with pitch as a binder and extruded as a rod. The rod is then baked to carbonize the pitch and graphitized by heating to about 2800°C. The electrodes used in a full-size arc steel furnace are up to 600 mm in diameter and 2·5 m long; they carry currents up to 85000 A, reach temperatures over 2000°C along the greater part of their length (and much higher at the tip where the arc forms) and are withdrawn from the furnace whilst still at high temperature. As carbon is consumed at the tip, the electrode is advanced and new lengths of rod are joined to the end of the old one. Full-sized electrodes have not yet been made from coal extracts and some of the following remarks are based on experience with electrodes made from petroleum feedstocks.

Table I shows some of the properties of single crystal graphite, which are highly anisotropic because of the layered structure. In an electrode containing graphite crystallites, the orientation of the crystallites will clearly influence the properties of the electrode, although there are other factors involved, such as the porosity and the nature of the binder. Two examples will be taken of properties of an electrode which are of importance in practice. Firstly, because of the very high electric currents passing through it, a low value of electrical resistance is required in the axial direction, and secondly, the electrode should be capable of withstanding the thermal shocks involved on withdrawal from the furnace.

Low electrical resistance in the axial direction is achieved by the use of a solvent extract of a high-rank coal. The coke derived from the extract has a lamellar texture ('needle coke') and the fragments produced on grinding tend to be elongated with the layer planes orientated roughly parallel to the long direction. During extrusion, such fragments become preferentially orientated with their long axes parallel to the extrusion direction, that is to say with the $c$ axis perpendicular to the axis of the electrode. X-ray analysis shows that

*Table I. Some physical properties of single-crystal graphite.*

|  | $a$ axis direction | $c$ axis direction |
|---|:---:|:---:|
| Carbon-carbon distance (nm) | 0·142 | 0·335 |
| Electrical resistivity (ohm.m) | $4 \times 10^{-7}$ | $2 \times 10^{-3}$ |
| Thermal expansion coefficient ($K^{-1}$) | $-1 \times 10^{-6}$ | $28 \times 10^{-6}$ |
| Thermal conductivity ($Wm^{-1}K^{-1}$) | $3 \times 10^{3}$ | 10 |
| Elastic modulus ($MNm^{-2}$) | $1 \times 10^{6}$ | $3 \times 10^{4}$ |

there is a distribution of orientations over the whole angular range from parallel to perpendicular, but up to six times as many crystallites may be in the 'right' direction as in the 'wrong' one.

In view of the very high degree of anisotropy of electrical resistivity in single crystal graphite, it may seem surprising that the ratio of resistivities perpendicular and parallel to the axis of an electrode is not greater than about three. Effectively, conduction *within* a crystallite is entirely in the $a$ direction, and there must be adequate conduction paths *between* neighbouring crystallites through the binder in both the perpendicular and parallel directions.

The factors of importance in relation to thermal shock resistance are not so clearly defined. When an electrode at, say, 2000°C is drawn out of the furnace, the outside surface cools rapidly and a steep temperature gradient will build up between the inside and the outside. This will generate tensile strain in the outer layer and a crack will be initiated if the strain exceeds the fracture strain, that is the ratio of tensile strength $(S)$ to elastic modulus $(E)$. The thermal strain is proportional to the coefficient of thermal expansion $(\alpha)$ and inversely proportional to the thermal conductivity $(k)$, so $kS/E\alpha$ is a possible measure of the resistance to crack initiation. If heat transfer from the surface is fast enough, however, the thermal conductivity should be omitted from the expression.

Once initiated, a crack will continue to propagate until there is insufficient thermal strain energy available to compensate for the energy expended in forming new crack surface. The size of crack will be inversely related to the fracture energy $(W)$ and directly related to the elastic energy stored at the instant of crack initiation, which is proportional to $S^2/E$, and $WE/S^2$ can therefore be taken as a measure of the resistance to crack propagation. Sato *et al.* (1974) have analysed the thermal stress fracture of a graphite electrode theoretically and have found that the most likely failure is by a crack propagating along a radius of the cylinder, in agreement with observations on an experimental furnace; the most important properties of the electrode in determining the thermal shock resistance should therefore be those measured in a transverse direction.

It is not yet established which (if either) of the expressions $S/E\alpha$ or $WE/S^2$ is the right one to use, but both of them show some correlation with the results of shock tests, provided that values of the parameters measured in a transverse direction are used (ambient temperature values have been used for lack of high temperature ones). The fact that both expressions correlate might seem surprising when one contains a factor $S/E$ and the other contains $E/S$, but the fracture strain $S/E$ does not appear to vary greatly between samples and therefore the predominant factors are $1/\alpha$ and $W/S$.

Nothing is known about the anisotropy of $W/S$ for crystalline graphite but Table I shows that the coefficient of thermal expansion $\alpha$ is large in the $c$ direction and very small in the $a$ direction. Consequently, the preferential orientation, which reduces the electrical resistivity in the axial direction of the electrode, tends to decrease the value of $1/\alpha$ in the transverse direction (and hence reduces the resistance to crack initiation). Measurements on electrodes indicate that the anisotropy of $W/S$ would have a similar effect to that of $1/\alpha$. As a compromise, Moore *et al.* (1974) recommend that a relatively coarse coke be used; the aspect ratios are reduced and the particles are therefore less well aligned, giving an advantageous reduction in $\alpha_{transverse}$ which outweighs the somewhat increased axial electrical resistance.

During the working life of an electrode, the surface becomes progressively affected by oxidation and special coatings have sometimes been commercially applied in an attempt to minimize this effect. These are not always a success and increased losses of material may be observed near the tip of the electrode. The reason for this could be that, in the ordinary course of events, oxidation reduces both $S$ and $E$, but $E$ is reduced much more than $S$. Hence $S/E$ increases, as does the resistance to crack initiation with the life of the electrode, that is towards the tip. The application of a coating may have inhibited this increase in resistance, leading to more spalling. The conclusion would be reversed, however, if propagation rather than initiation was the vital step.

## V. CONCLUSION

Metallurgical coke has been an essential coal-based raw material in the manufacture of iron and steel for over 200 years and is likely to continue to be so (even when the best coking coals become very scarce) through the skilful blending of coals based on a better understanding of the structural factors which affect the coke properties.

Considerable scope also exists for the application of these principles in the use of other coal-based carbonaceous materials, such as cokes derived from coal extracts, which could replace petroleum-based carbons in the manufacture of graphite electrodes for arc steel furnaces.

## REFERENCES

Diamond, R. (1960). *Phil. Trans. R. Soc.* **252,** 193-223.
Franklin, R. E. (1951). *Proc. R. Soc.* A **209,** 196-218.
Gryaznov, N. S. and Kopeliovich, L. V. (1975). *Coke Chem. USSR* **7,** 4-7.
Moore, R. W., Smith, M. J. and Ince, A. (1974). 4th SCI International Carbon and Graphite Conference. Extended abstracts, pp. 216-217.
Patrick, J. W., Reynolds, M. J. and Shaw, F. H. (1973). *Fuel, Lond.* **52,** 198-204.
Sato, S., Sato, K., Nikaido, M. and Kon, J. (1974). *Carbon* **12,** 555-571.
van Krevelen, D. W. (1961). "Coal. Typology-chemistry-physics-constitution." Elsevier, Amsterdam.
White, J. L. (1975). *Prog., Solid State Chem.* **9,** 59-104.

# 5 Ultrastructural Analysis of Coal Derivatives by Electron Microscopy

## G. R. MILLWARD

*Edward Davies Chemical Laboratories, University College of Wales, Aberystwyth*

## I. INTRODUCTION

Although it was as long ago as 1956 that atomic lattice images of crystalline solids were first reported (Menter, 1956), it was not until much later that improvements in the performance of electron microscopes enabled investigators to obtain lattice 'fringe' images that could be associated with 'edge-on' projection views of stacks of graphite layers (Heidenreich *et al.,* 1968; Fourdeux *et al.*, 1969; Hugo *et al.*, 1970; Johnson, 1970; Marsh *et al.*, 1971). An example illustrating the "onion-skin-like" arrangement of graphitic layers in heat-treated carbon blacks is shown in Fig. 1. These developments opened the way to a ready means of obtaining direct insight into the molecular architecture of graphitic carbons (see Ban, 1972, and Millward and Jefferson, 1978, for reviews), without undergoing the bulk, spatial averaging processes, which are necessarily incurred with X-ray diffraction techniques, etc.

Much of the work reported in the literature on lattice-imaging studies of graphitic carbons has, unfortunately, been influenced detrimentally by electron optical lens aberrations; it cannot, necessarily, be assumed that the lattice 'fringes' in the micrographs correspond precisely to a projection of the graphite layers in the specimen. Such problems of image formation in relation to carbons are now fully understood (Thon, 1966; Johnson and Crawford, 1973; Millward and Thomas, 1976; Jefferson *et al.*, 1976; Millward and Jefferson, 1978), and optimum conditions for imaging have been specified (Crawford and Marsh, 1977; Millward and Thomas, 1978). In the case of graphitic materials where the variation of interlayer spacing is large, or the crystallite size (number of regularly stacked layers) is small, it is only the very recent generation of electron microscopes (e.g. Jeol 100/200 CX, Philips EM400 or Siemens CT150) that have sufficiently good characteristics to yield reliable and unambiguous results (Millward and Thomas, 1978).

Notwithstanding these reservations, the lattice 'fringe' images available in the literature can be used to characterize, in a qualitative way, many of the structural features found in various carbons subjected to a wide range of graphitizing conditions. In particular, it is possible to derive information about the environment and relative orientation of graphite sheets, or groups of sheets, and about crystallite size and the degree of stacking order, provided that the limitations imposed by the imperfect nature of the imaging properties of the electron microscope are appreciated.

The first electron microscope investigations of significance into the

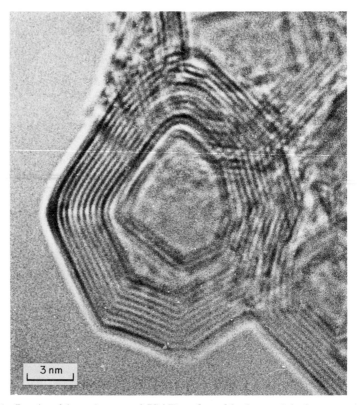

*Fig. 1. Lattice-fringe image of ISAF carbon black particle heat-treated at 2600°C.*

ultramicrostructure of heat-treated coals and coal derivatives were reported by Evans *et al.* (1972). After sections devoted to electron microscope technique and the problems of image formation, there will follow in this chapter an indication of the type of information which has become available from lattice image studies of a range of heat-treated, coal-based materials.


## II. ELECTRON MICROSCOPY

### Experimental techniques

#### 1. Specimen preparation

Because of the high degree of interaction of electrons with matter, at the accelerating potentials common in conventional high-resolution electron microscopes (about 100KeV), specimens must be very thin. Generally, the thickness parallel to the electron beam must be less than 100 nm if the specimen is to exhibit any sensible degree of transparency and, in the case of graphitic carbons, very much thinner if the effects of multiple scattering are to be absent from the image (Jefferson *et al.*, 1976).

The simplest, and most widely used method of producing thin specimens is to grind the material, for example, with an agate mortar and pestle. The resulting powder can be mounted on the electron microscope specimen grid either by dusting directly, or from suspension in ethanol, acetone or an inert liquid. In order that very small fragments can be supported, the standard copper mesh specimen grid (the mesh usually has holes about 100 $\mu$m square) is covered with a carbon-coated, thin plastic film which is permeated throughout by a network of small holes. An excellent method of preparing such films has been described by Fukami and Adachi (1965). Sufficiently thin regions of a fragment lying over a hole in the supporting film are chosen for microscopy.

It follows that high resolution electron microscopy is a highly selective technique, and results should be considered within this context. The investigator is behoven to specify when quoted results are not generally typical of the specimen examined.

A criticism which can be levelled at grinding methods of specimen preparation is that all spatial relationships are lost between the fragment being examined and the bulk material from which it was derived. Further, the existence of preferential cleavage planes in the bulk material may result in fragments so shaped that they rarely present certain structural features in the correct orientation in the specimen plane of the electron microscope. (This is often true of the more highly ordered graphitic carbons which yield flat fragments, which tend to lie with the planes of the graphite sheets at right angles to the viewing axis.)

Ultrathin-sectioning methods (ultramicrotomy), which have been developed largely for electron microscopy applied in biological studies, do not suffer from these disadvantages, and selected orientations of the specimen can be chosen. Considerable manipulative skill is required for this method, but it has been applied successfully, if not with ease, for preparing thin sections (longitudinal and transverse) of carbon fibres (Harling, 1971; Mencik *et al.*, 1975).

## 2. Electron microscope operation

Most high resolution instruments that have been available on a commercial basis during the past ten years, have incorporated sufficiently good mechanical and electronic stability and quality of lens design, for lattice periodicities at least as small as the primary graphite (0002) spacing (0·34 nm) to be readily resolved. In order to optimize the lattice 'fringe' contrast in the visible or recorded image, the operational conditions of the illumination system have to be chosen with care. The gun should be operated to yield as small a beam current as possible, consistent with the obvious necessity for recording the image on a photographic emulsion at the desired magnification and in a reasonable (less than 10s) exposure time. An excessive beam current results in a wider than necessary spread in the range of energies of the source electrons (Boersch, 1954); this, coupled with chromatic aberration of the electron-optical lenses causes loss of contrast for the smaller lattice spacings (Hanzen and Trepte, 1971). A similar diminution in high resolution contrast occurs if the solid angle of illumination at the specimen is too large (Hanzen, 1971; Frank, 1973). Normally, for high resolution work, the condenser system is operated so that the electrons are fully focused to a spot on the specimen (*critical illumination*) as described by Barnett (1975), and the angle of illumination is governed by the diameter of the exit aperture of the condenser lens system, together with the converging properties of the prefield of the objective lens in which the specimen is usually situated. In practice, for resolution of the 0·34 nm periodicity associated with graphitic substances, an illumination solid half-angle of $10^{-3}$ rad, or less, is satisfactory. In the case of a Philips EM300 electron microscope (high resolution stage and lens) the prefield factor is about 1·2 and the appropriate aperture to use in the second condenser lens (situated 130 mm from specimen) is one of diameter 200 $\mu$m (illumination half-angle 9·2 $\times$ $10^{-4}$ rad).

In many types of lattice resolution studies, the choice of size of objective aperture is important (Cowley and Iijima, 1972). However, in the case of graphitic (0002) lattice imaging (viewing along the layers), provided that the aperture does not cut off the dominant (0000), (0002) and (000$\overline{2}$) diffracted beams, the aperture size is not critical; it is often omitted.

Perhaps the most problematic factor in operating an electron microscope for lattice resolution is the defocus condition of the objective lens. It has been recognized from the earliest days of lattice imaging applications (Menter,

1956) that optimum phase-contrast images are, in general, obtained when the objective lens is slightly underfocused (weak lens) from the true Gaussian focus. In fact there are many positions of defocus when a 'lattice-fringe' image (but not necessarily a true *structural image*) can be observed with good contrast (Johnson and Crawford, 1973; Jefferson *et al.*, 1976; Millward and Jefferson, 1978). It is, therefore, important that the operator is familiar with the now well-established reasons for this behaviour (see next section) in order that he can choose the optimum defocus for his particular purposes.

It may be necessary, in many cases, to take a series of photographs of the specimen, each photograph being recorded at a small increment in defocus from the previous one, so that the 'optimum' image can be chosen later. Alternatively, and more conveniently, the use of a fibre optics/TV display attachment (preferably with image intensifier) enables a more direct and simple judgement of the image to be made prior to it being recorded photographically. In both cases final assessment of the image can be made using optical diffraction analytical methods (Thon, 1966; Millward and Thomas, 1976; Millward and Jefferson, 1978), or equivalent computational methods.

## Theoretical considerations of image formation in the electron microscope

There are two basic steps to consider in any treatment of image formation, namely, interaction of the incident wave with the object, and transfer of the perturbed wave to the image plane, by the optical system.

### 1. Interaction of electrons with the object

If the specimen is very thin, the perturbations imposed on the incident wave are small and the specimen can be treated approximately as a *weak-phase/amplitude* object (Hanzen, 1971). For ultra-high resolution, lattice imaging of atomic detail, the dominant electron-scattering mechanism, related to image contrast, is of phase origin and the amplitude (absorption) factor can usually be ignored. As the thickness of the specimen (in the direction of the electron beam) becomes greater, the weak-phase-object approximation becomes invalid, multiple scattering dynamical effects become significant, and the image, even for a perfect optical system, does not necessarily look like the original object structure (O'Keefe, 1973). Multiple scattering as it affects (0002) lattice images of graphitic carbon, has been considered by Jefferson *et al.*, (1976), and it was concluded that only specimens thinner than 5 nm can be considered to satisfy, even qualitatively, the weak-phase-object approximation.

### 2. Object to image transfer

In order to understand the processes involved in the transfer of the perturbed wave from the object to the image plane, it is useful to refer to a simple diagrammatic illustration (Fig. 2) of the principles of image formation. (Although an electron microscope is a multi-lens system, it is the quality and aberration properties of the *objective* lens which dominate the image formation, so the important principles can be considered in terms of a single lens model.) In Fig. 2 the object is depicted as perturbing an incident plane-wave front to give a *primary* wave (unscattered electrons) and two symmetrically *diffracted* waves (scattered electrons). The situation therefore resembles the edge-on graphite scattering situation with (0000), (0002) and (000$\bar{2}$) diffracted beams. The waves are collected by the lens and brought to a focus in the back-focal plane of the lens (Fraunhofer diffraction pattern) and subsequently propagate forwards to interfere with each other in the image plane. In the case of an ideal weak-phase object the diffracted waves immediately after the object will be $\pi/2$ out of phase with the primary wave (Hanzen, 1971), and in the Gaussian image plane of a perfect optical system there will be no significant interference effects and, therefore, no image contrast. In order to promote optimum interference in the image plane it is necessary to induce an additional relative phase change of $\pi/2$ between the diffracted waves and the primary wave, and thereby promote the formation of an interference fringe image of the object structure (Born and Wolf, 1966; Lipson and Lipson, 1969; Hanzen, 1971). In an electron microscope, unavoidable phase changes in diffracted beams, relative to the primary beam, are induced by the spherical aberration of the objective lens; further

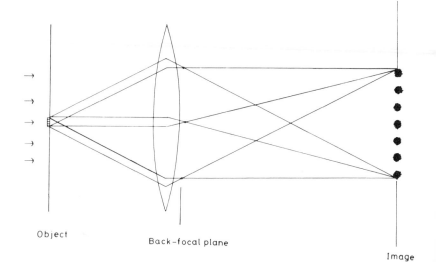

*Fig. 2. Schematic diagram illustrating the basic principles of phase-contrast image formation by a single lens.*

phase changes may be introduced by defocusing the lens. Such phase changes are related to the diffraction angle and by judicious choice of defocus (DF), to complement the spherical aberration factor $(C_s)$, the desired net $\pi/2$ instrumental phase change can, in certain restricted instances, be obtained. This is the basis of phase-contrast electron microscopy and has been fully treated mathematically elsewhere, (Hoppe, 1970; Hanzen, 1971; Cowley and Iijima, 1972; Misell, 1973) in terms of *linear transfer theory*.

One of the essential predictions of transfer theory is that for a large, regularly ordered, graphitic crystal, where electrons will be scattered at a single specific angle by the (0002) lattice, the image-fringe positions will be fixed but their contrast will vary periodically with objective lens defocus (positive, zero or negative contrast). Consequently there will be many positions of defocus which will yield a 'correct' image and aberration problems are not usually serious. However, in the case of a small crystal or an irregularly ordered lattice, where electrons will be scattered over a relatively wide continuous range of angles, it is only defocus positions very close to the so-called *Scherzer-focus* (Scherzer, 1949; Cowley and Iijima, 1972; Millward and Thomas, 1976; Millward and Jefferson, 1978) which will be expected to yield true structural images, and only then if the aberration parameters (notably spherical aberration) are sufficiently good.

Images exhibiting spurious fringes as a result of incomplete information transfer in the electron microscope have been demonstrated for a variety of graphitized or partially graphitized carbons, either experimentally (Johnson and Crawford, 1973; Crawford and Marsh, 1977) or by theoretical computer image simulation studies (Jefferson et al., 1976), or by a combination of the two (Iijima, 1977; Millward and Jefferson, 1978; Millward and Thomas, 1978; Millward et al., 1978). An example is shown in Fig. 3 where a twin-layered graphitic crystal is shown imaged at differing focal levels and compared with computer simulated images based upon transfer theory. The most appropriate experimental image is Fig. 3b corresponding to a 'near-Scherzer' focus. Reversal of contrast and additional spurious fringes appear for inappropriate transfer conditions (Figs 3a, c). The detrimental influence of inadequate transfer properties can also be seen in a computer image-simulation study (Millward and Jefferson, 1978; Millward and Thomas, 1978) of a model graphitic structure exhibiting irregular stacking of the sheets (spacings 0·34 nm to 1·0 nm) (Fig. 4). Comparison of the ideal image of the model with variously defocused images appropriate to a Philips EM300 electron microscope ($C_s = 1·6$ mm), shows that there is no fully satisfactory choice of focus available. The 'near-Scherzer' focus condition (DF $= 92·5$ nm) is, perhaps, the most acceptable. In order to obtain better images of such disordered structures it is necessary to improve the 'near-Scherzer' transfer condition by using lenses with lower $C_s$ values (see Fig. 4, lower two images), or by operating at higher accelerating potential (Millward and Thomas, 1978).

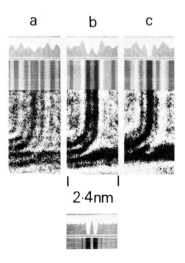

Fig. 3. *Comparison of experimental and computer-simulated images of a pair of graphite sheets. The contrast convention* (print contrast) *is that a* black *fringe indicates the position of a layer in a correct image. In the model the sheets are assumed to be 0·34 nm apart and extend for 0·5 nm in the direction of the incident* (100 KeV) *electron beam, (weak-phase object). Objective aperture spatial frequency cut off is 5 nm⁻¹ (resolution limit 0·2 nm). (a) $C_s = 1·6$ mm, 45 nm underfocus, (b) $C_s = 1·6$ mm, 92·5 nm underfocus, (c) $C_s = 1·6$ mm, 140·0 nm underfocus, (d) 'ideal' image assuming perfect optics.*

## III. DIRECT OBSERVATION OF THE DEVELOPMENT OF GRAPHITIC ORDER IN COALS AND COAL DERIVATIVES AS A RESULT OF HEAT TREATMENT

The development of graphitic order in different classes of coal, as the temperature of heat-treatment is increased, has been described in Chapter 4 in terms of information derived from X-ray diffraction analysis and optical microscopy. Direct evidence about the arrangements of the graphite layers, obtained by electron-microscope lattice imaging, is an obviously attractive complementary technique.

When untreated coals are examined in the electron-microscope, no evidence is found of 'fringes' which could be associated with layered molecules. The images (e.g. Fig. 5) resemble those of evaporated carbon films (Thon, 1966), and the observed detail is as much a function of the transfer properties of the electron microscope as of the structure of the

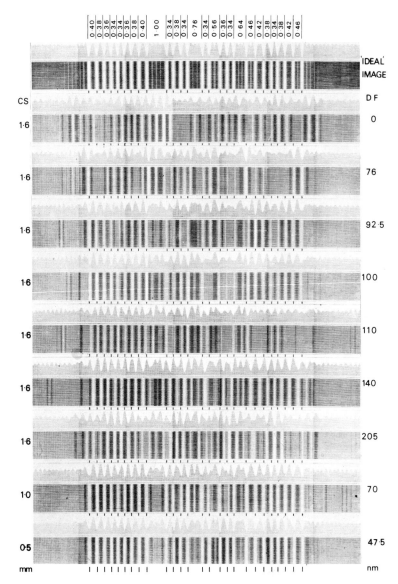

*Fig. 4. Computer-simulated images of a non-periodic model array of graphite layers. The contrast convention adopted* (plate contrast) *depicts a layer as a* white *fringe in a correct image. The model extends for 0·5 nm in the direction of the incident (100 KeV) electron beam and is good approximation to a weak-phase object. The positions and spacing of the layers are indicated in relation to the idealized image, calculated assuming perfect optics. Objective aperture spatial frequency cut off is 5 nm⁻¹ (resolution limit 0·2 nm).*

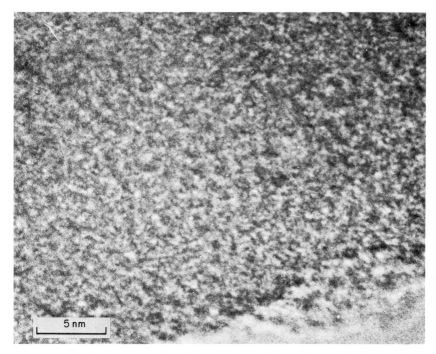

*Fig. 5. High-resolution phase-contrast image of raw Ogilvie coal.*

specimen (Thon, 1966; Millward and Jefferson, 1978). In view of the presently accepted, generalized structure of coals (see Chapter 2), where planar aromatic layers are small, and where there is very little specific spatial interrelationship between neighbouring layers, the lack of fringe structure in electron micrographs of raw coal is not very surprising, especially if one considers the damage that organic molecules usually undergo when subjected to the intense radiation of an electron microscope used at high magnification (Glaeser, 1971).

The study of the incipient stages of graphite layer growth and assembly (heat treatments below 1000°C) by electron microscopy has been hampered by the defective nature of images generally obtained from disordered molecular arrangements, (see Section II, *Object to image transfer*). Nevertheless, by optimizing, very precisely, the imaging conditions to a 'near-Scherzer' focus position it is possible to derive useful, if limited, data about graphitic structure developed at relatively low temperatures. Thus Crawford and Marsh (1977) were able to describe, in qualitative terms, progressive stages of growth and alignment of molecular layers in Orgreave lean coal-tar pitch heated to temperatures between 477°C and 875°C under nitrogen. They suggest that the material treated at 477°C contains small, non-aligned layers isolated from larger layers, the latter exhibiting some degree of preferential orientation. By 672°C many of the small layers have become attached to the larger layers and the system, overall, is more distorted. During further heat treatment to 875°C the distorted layers anneal and become larger and more planar. Until results of this type are obtained using instruments with better electron-optical properties than have been available until very recently, we must retain some reservations about their precise interpretation; nevertheless, the observations do illustrate the potential of the method.

As the temperature of heat treatment of coal increases so does the ordering of the fringes in the corresponding lattice images. Thus Fig. 6 is a typical electron micrograph of a cabonized coal extract (prepared from 301 (a) coal) that has been treated at a temperature of 1300°C. The fringe pattern in this micrograph is significantly different from that observed when the same coal extract is heated to 2480°C (Fig. 7). In the former instance (1300°C) the alignment and ordering of the fringes is much poorer than in the latter case. Oberlin and Terriere (1975) have demonstrated similar improvements in lattice-fringe ordering as the temperature of heat-treatment of anthracites becomes higher.

Before leaving this section we should consider how the images of Figs 6 and 7 might be affected by instrumental transfer problems. In the case of Fig. 7 the lattice fringes are particularly well ordered over extended distances, and the image is probably a very good representation of the structure (see Section II, *Object to image transfer*). We cannot, however, be certain whether it is the black or white fringes which correspond to graphite layers (positive or negative transfer), and some errors will inevitably be found in regions where the lattice is defective. In Fig. 6 the stacking and relative alignment of the fringes is not nearly so good, and this certainly reflects the less well organized graphitic structure in the 1300°C sample. However, because of this disorder, which will cause electrons to be scattered over a wide range of angles, the instrumental transfer factors are likely to have influenced the image so that it contains misplaced and spurious fringes and reversals in fringe contrast (Millward and Jefferson, 1978; Millward and Thomas, 1978); it could well indicate a better organized structure than is in fact the case (Crawford and Marsh, 1977).

## IV. VARIATIONS IN GRAPHITIC STUCTURE OF HEAT-TREATED COALS ACCORDING TO ORIGIN

Examples illustrating different forms of graphitic growth, which can be associated with different coals, are shown in Figs 8 and 9.

Figure 8 is an electron micrograph of a typical fragment of low-rank (702) Askern coal heat-treated at 1000°C. The 'layer-fringes' are randomly

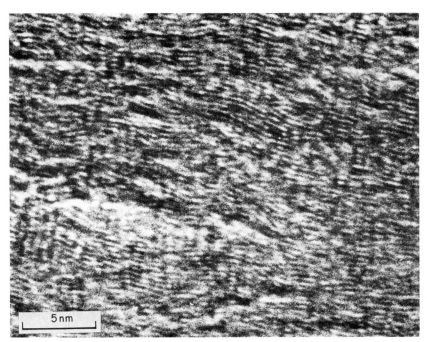

*Fig. 6. Lattice-fringe image of 301 (a) coal extract heat-treated at 1300°C.*

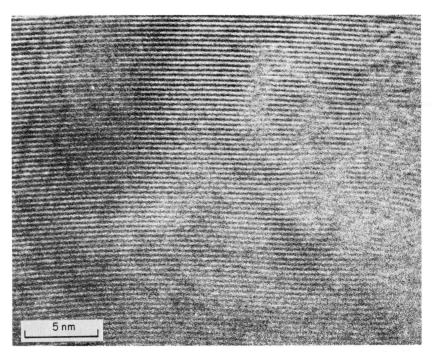

*Fig. 7. Lattice-fringe image of 301 (a) coal extract heat-treated at 2480°C.*

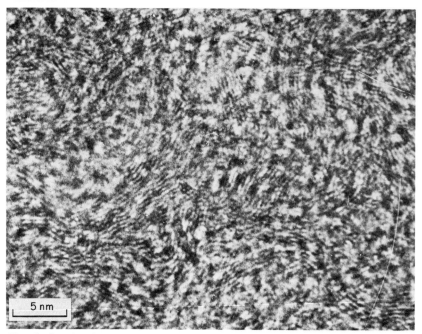

*Fig. 8. Lattice-fringe image of Askern (702) coal heat-treated at 1000°C.*

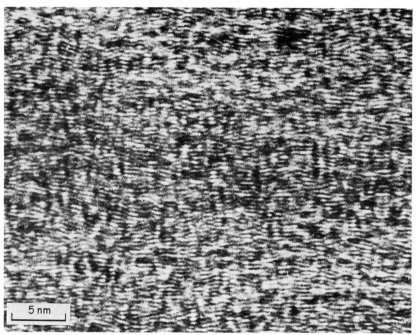

*Fig. 9. Lattice-fringe image of Ogmore (204) coal heat-treated at 1000°C.*

orientated and not well ordered; this image is typical of the so-called
non-graphitizing carbons (Jenkins *et al.*, 1972; Oberlin *et al.*, 1975; Ban *et al.*, 1975) heated at this temperature. In such materials considerable
three-dimensional cross-linking develops between the small layers at low
temperatures. The resulting rigidly cross-linked, randomized structure is not
amenable to major realignment of layers at higher temperatures (see Chapter
4) so that, although the ordering of layers might improve, the basic
randomness of the structure persists (e.g. Fig. 10).

In the case of a high-rank (204) Ogmore coal (heated at 1000°C), the
'layer-fringes' are commonly much better ordered and exhibit a marked
habit of preferential orientation (Fig. 9). Such images are typical of
graphitizing carbons (Oberlin *et al.*, 1975) where the cross-linking developed
at low temperatures is much less intensive than for non-graphitizing carbons,
and realignment of layers is more easily permitted at higher treatment
temperatures (e.g. Fig. 7).

These results should not be taken too literally; they portray a simplified
picture of the overall structure of heat-treated coals. Coals are usually quite
heterogeneous in structure and composition, their classification being determined by essentially averaging techniques (see Chapters 1 and 2). The
'fringe' images shown in Figs 8 and 9 are certainly representative of fragments

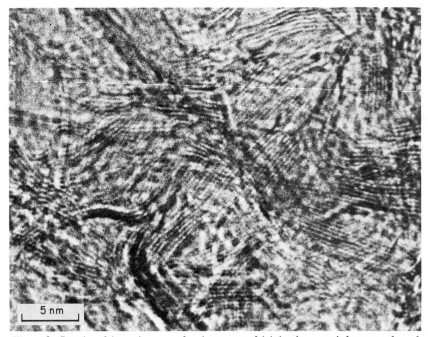

5 nm

*Fig. 10. Lattice-fringe image of a 'non-graphitizing' material, p-terphenyl, heat-treated at 3000°C.*

from their respective coal types, but this does not exclude the presence of minor components of differing graphitic texture in either case. Coals of intermediate rank classification, between Askern (702) and Ogmore (204), exhibit much more heterogeneity of structure, and their characterization by electron microscopy has yet to be effectively explored.

Anthracites possess rather different graphitizing properties to other coals (see Chapter 4), because they combine, in the raw, natural state, a relatively high degree of parallel layer alignment imposed by past geological conditions, together with considerable three-dimensional cross-linking. Oberlin and Terriere (1975) have demonstrated, in the electron microscope, the alignment of layers in anthracites treated at various temperatures (e.g. Figs 11 and 12). They conclude that the structure of anthracites is basically an anisotropic, foam-like texture, and that the preferred layer orientation exists because the pores (and surrounding walls) have been flattened along the bedding plane.

Very often, depending on the origin and impurity content of the raw material, anomalous graphitic structures are found which are quite different from the bulk (Evans et al., 1972; Oberlin and Terriere, 1975; Johnson et al., 1975). Electron microscopy is particularly useful, in such circumstances,

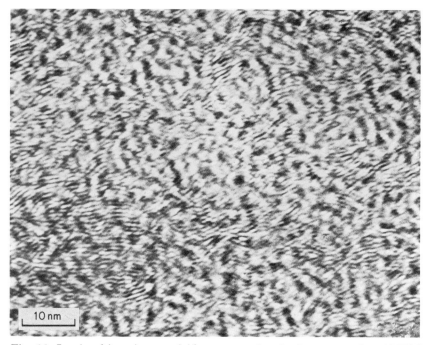

*Fig. 11. Lattice-fringe image of Abernant anthracite heat-treated at 1000°C (Oberlin and Terriere, 1975).*

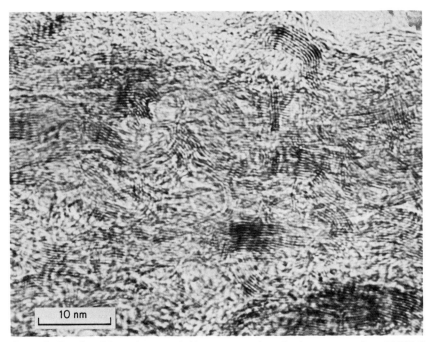

*Fig. 12. Lattice-fringe image of Abernant anthracite heat-treated at 2000°C
(Oberlin and Terriere, 1975).*

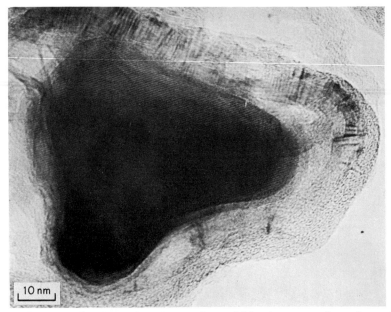

*Fig. 13. Example of catalytic-induced graphitization around an impurity
particle in Onllwyn anthracite heat-treated at 1370°C. (Evans et al., 1972).*

enabling isolated rare structural features to be observed. In most other analytical techniques the features would be lost in the averaging process, or be in too small a quantity to be detected. Examples of impurity-induced, catalytic graphitization in anthracites have been reported by Oberlin and Terriere (1975), and by Evans *et al.* (1972), who showed, for the temperature of treatment employed, unexpectedly well ordered graphite surrounding impurity particles (Fig. 13).

## V. ELECTRON MICROSCOPY OF SPECIFIC PRODUCTS FROM COAL

### Carbon fibres

Lattice imaging in the electron microscope has been used extensively for characterizing ultrastructure in carbon fibres produced from preformed plastic fibres (Fourdeux *et al.*, 1969; Hugo *et al.*, 1970; Crawford and Johnson, 1971; Johnson and Crawford, 1973). Carbon fibres have also been produced at the National Coal Board Research Establishment, Stoke Orchard (see Chapter 9), by suitable extrusion of coal extract followed by

*Fig. 14. Lattice-fringe image of low strength/modulus carbon fibre prepared from coal extract* ($E_m = 41\ GNm^{-2}$, $T_s = 0.66\ GNm^{-2}$, *see text*).

carbonization (300-1000°C) and strain-graphitization (at 2700°C) stages (Ladner *et al.*, 1976; Ladner and Jorro, 1976). Differences in the ultra-structure of these NCB fibres have been observed in the electron microscope, typical images being shown in Figs 14-16. Figure 14 refers to a low strength/modulus fibre (tensile strength, $T_s = 0.66$ GNm$^{-2}$; Young's modulus, $E_m = 41$ GNm$^{-2}$); the lack of development of significant graphitic order is apparent. However, strain-graphitization at high temperature induces the growth of assemblies of well ordered graphitic sheets orientated along the fibre axis (Ladner and Jorro, 1976), and this is well illustrated by the lattice images of Fig. 15 (fibres where $T_s = 0.94$ GNm$^{-2}$, $E_m = 150$ GNm$^{-2}$) and Fig. 16 (fibres where $T_s = 2.07$ GNm$^{-2}$; $E_m = 377$ GNm$^{-2}$). The lower strength/modulus values of the fibres associated with Fig. 15, compared with those associated with Fig. 16, correlate with a poorer orientation of the graphitic crystallites, and a higher incidence of very small assemblies of layers (one, two or three only); the latter small groups of sheets are often inclined at large angles to the fibre axis. This correlation of ultrastructure with fibre strength modulus characteristics is obviously of significance. Nevertheless it can only be a part of the story, and the effects of other factors, such as defects on a macro-scale, should not be ignored (Reynolds, 1973).

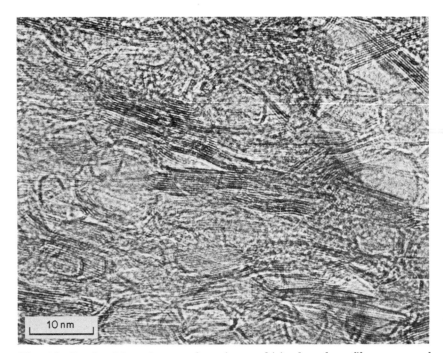

*Fig. 15. Lattice-fringe image of strain-graphitized carbon fibre prepared from coal extract* ($E_m = 150$ *GNm*$^{-2}$, $T_s = 0.94$ *GNm*$^{-1}$, *see text*).

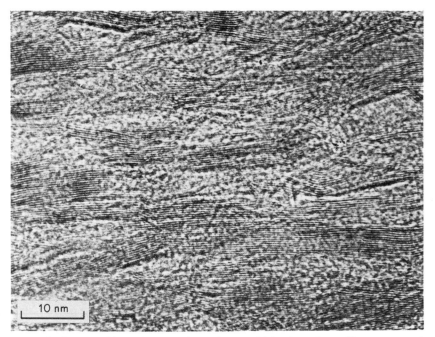

*Fig. 16. Lattice-fringe image of strain-graphitized carbon fibre prepared from coal extract* ($E_m = 377\ GNm^{-2}$, $T_s = 2\cdot07\ GNm^{-2}$, *see text*).

## Activated carbon

The preparation of active carbon from anthracite is described in Chapter 9, the final product consisting of a system of interconnected open pores surrounded by carbon walls. High resolution electron microscopy has been used to show (Stoeckli and Huber, 1977; Millward *et al.*, 1978; Millward and Jefferson, 1978) that, in some preparations, the structure of the material is so expanded that the walls around the pores often consist of only single sheets of carbon atoms (see Fig. 17).

The problem of whether or not it is feasible to observe, edge-on, single sheets of carbon atoms in the electron microscope has been considered by Millward *et al.* (1978), their conclusions being that it is, provided that the instrument is operated near to the Scherzer focus.

## VI. CONCLUSIONS

I have presented here the results of direct ultramicrostructural analysis of some graphitized and partially graphitized carbons utilizing phase-contrast,

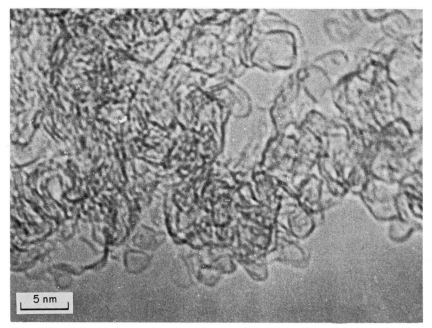

*Fig. 17. High resolution image of strongly activated anthracite, showing pores surrounded by single layers of carbon atoms. (Courtesy Prof. H. F. Stoeckli and Dr D. Crawford).*

lattice-imaging electron microscopy. The limitations of such images, imposed by electron-optical aberrations, have been demonstrated by theoretical image calculations, and experimentally, and the importance of choosing transfer conditions appropriate to the type of specimen being studied has been stressed. In order for true structural images of graphitic substances to be obtained, the objective lens aberration factors are required to be of a sufficiently high standard, and the instrument operated near the Scherzer focus position.

It is only very recently that instruments with adequate capability have become available commercially (Siemens CT150, Jeol 100/200 CX, Philips EM400). Consequently the previously unavoidable electron optical image defects in lattice images of graphitic materials should present few problems in the future.

# REFERENCES

Ban, L. L. (1972). *In* 'Surface and Defect Properties of Solids.' (M. W. Roberts and J. M. Thomas, Eds.) Vol. 1, pp. 54-94. Chemical Society, London.

Ban, L. L., Crawford, D. and Marsh, M. (1975). *J. Appl. Crystallogr* **8,** 415-420.

Barnett, M. E. (1975). *J. Microsc.* **102,** 1-28.

Boersch, H. (1954). *Z. Phys.* **139,** 115-146.

Born, M. and Wolf, E. (1966). 'Principles of Optics.' (5th ed.) Pergamon Press, Oxford.

Cowley, J. M. and Iijima, S. (1972). *Z. Naturforsch.* **27a,** 445-451.

Crawford, D. and Johnson, D. J. (1971). *J. Microsc.* **94,** 51-62.

Crawford, D. and Marsh, K. (1977). *J. Microsc.* **109,** 145-152.

Evans, E. Ll., Jenkins, J. Ll. and Thomas, J. M. (1972). *Carbon* **10,** 637-642.

Fourdeux, A., Herinckx, G., Perret, R. and Ruland, W. (1969). *Comptes Rendus* C **269** 1597-1600.

Frank, J. (1973). *Optik* **38,** 519-536.

Fukami, A. and Adachi, K. (1965). *J. Electronmicrosc.* **14,** 112-118.

Glaeser, R. M. (1971). *J. Ultrastructure Res.* **36,** 466-482.

Hanzen, K. J. (1971). *In* 'Advances in Optical and Electron Microscopy.' (R. Barer and V. E. Cosslett, Eds.) Vol. 4, pp. 1-84. Academic Press, London and New York.

Hanzen, K. J. and Trepte, L. (1971). *Optik* **32,** 519-538.

Harling, D. F. (1971). *Jeol News* **8,** 22-26.

Heidenreich, R. D., Hess, W. M. and Ban, L. L. (1968). *J. Appl. Crystallogr.* **1,** 1-19.

Hoppe, W. (1970). *Acta Crystallogr.* A **26,** 414-426.

Hugo, J. A., Phillips, V. A. and Roberts, B. W. (1970). *Nature, Lond.* **226,** 144.

Iijima, S. (1977). Proc. 35th Ann. EMSA meeting, p. 194.

Jefferson, D. A., Millward, G. R. and Thomas, J. M. (1976). *Acta Crystallogr.* A **32,** 823-829.

Jenkins, G. M., Kawamura, D. and Ban, L. L. (1972). *Proc. R. Soc.* A **327,** 501-517.

Johnson, D. J. (1970). *Nature, Lond.* **226,** 750-751.

Johnson, D. J. and Crawford, D. (1973). *J. Microsc.* **98,** 313-324.

Johnson, D. J., Tomizuka, J. and Watanabe, O. (1975). *Carbon* **13,** 321-325.

Ladner, W. R. and Jorro, M. A. A. (1976). *In* 'Proceedings of the 4th London Int. Carbon and Graphite Conference, 1974.' pp. 287-303. Soc. of Chem. Industry, London.

Ladner, W. R., Jorro, M. A. A. and Randall, T. D. (1976). *Carbon* **14,** 219-224.

Lipson, S. G. and Lipson, H. (1969). 'Optical Physics.' Cambridge University Press.

Marsh, P. A., Voet, A., Mullens, T. J. and Price, L. D. (1971). *Carbon* **9,** 797-805.

Mencik, Z., Plummer, H. K. Jr. and Barosiewicz, L. (1975). *Carbon* **13,** 417-420.

Menter, J. W. (1956). *Proc. R. Soc.* A **236,** 119-135.

Millward, G. R. and Jefferson, D. A. (1978) *In* 'Chemistry and Physics of Carbon.' (P. L. Walker, Jr. and P. A. Thrower, Eds.). Vol. 14. Marcel Dekker, Inc., New York and Basel (in press).

Millward, G. R. and Thomas, J. M. (1976). *In* 'Proceedings of 4th London Int. Carbon and Graphite Conference, 1974.' pp. 492-497. Soc. of Chem. Industry, London.

Millward, G. R. and Thomas, J. M. (1978). *Carbon* (in press.)

Millward, G. R., Jefferson, D. A. and Thomas, J. M. (1978). *J. Microsc.* **113,** 1-13.

Misell, D. L. (1973). *In* 'Advances in Electronics and Electron Physics.' Vol. 32, pp. 63-191. Academic Press, New York and London.

Oberlin, A. and Terriere, G. (1975). *Carbon* **13,** 367-376.

Oberlin, A., Terriere, G. and Boulmier, J. L. (1975). *Tanso* **80**, 29-42.

O'Keefe, M. A. (1973). *Acta Crystallogr.* A **29**, 389-401.

Reynolds, W. N. (1973). *In* 'Chemistry and Physics of Carbon.' (P. L. Walker, Jr. and P. A. Thrower, Eds). Vol. 11, pp. 1-67. Marcel Dekker, Inc., New York and Basel.

Scherzer, O. (1949). *J. Appl. Phys.* **20**, 20-29.

Stoeckli, H. and Huber, U. (1977). *Agents and Actions* **7**, 411-420.

Thon, F. (1966). *Z. Naturforsch.* **21a**, 476-478.

# 6  The Combustion of Coal

## A. D. DAINTON

*Director, Coal Research Establishment, National Coal Board*

## I. INTRODUCTION

Combustion of coal is by far the most important method of coal consumption in this country. In 1975, for example, 120 million tonnes of coal were consumed. Of this figure, about 20 million tonnes were used in the manufacture of coke and chemical products. The remaining 100 million tonnes were burned—73 million tonnes in electricity generation, the rest in steam-raising and space heating.

The pattern of coal utilization will undoubtedly change over the next 25 years. Gasification and liquefaction of coal will become important, but it is probable that combustion will remain the biggest single method of use. In this chapter, some basic physical principles of combustion will be discussed, briefly illustrated by reference to industrial stoking methods in common use, but the real emphasis will be placed upon the relatively new technique of fluidized bed combustion, which is a subject of world-wide interest and is likely to become widely used in the future.

## II. METHODS OF COAL COMBUSTION

### Up-draught combustion

Combustion of coal may be defined as the oxidation of carbon and hydrocarbons to carbon dioxide and water, with the accompanying release of heat. One of the simplest ways of achieving this is shown in Fig. 1 which represents a vessel containing the lumps of solid fuel in a bed supported by a grate. Provision is made for the supply of primary air beneath the bed and secondary air above it, and there is a connection to a flue in order to provide a draught. This simple principle has done sterling service in a number of applications ranging from the domestic fire to the blast furnace. It is common knowledge that a fire of this type is ignited at the bottom. The flame front then spreads upward until the whole bed is incandescent. The system is generally termed 'up-draught' combustion, but the important technical feature is that the flame front travels in the same direction as the primary air.

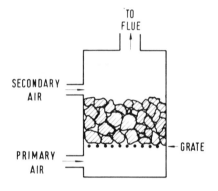

*Fig. 1. Simple fire.*

It is at this point that simplicity ceases to be a virtue and begins to cause difficulties. The majority of British coals contain between 30 and 40% of volatile matter, including a quantity of tar which is evolved at around 450°C. In this simple combustion method, heat is transferred ahead of the flame front by radiation and convection, and it distils the tars at temperatures below their ignition temperature. The purpose of the secondary air is to burn the volatile matter, but whereas in simple appliances it is not difficult to supply the necessary air, there is rarely sufficient turbulence to mix it with the volatiles, and the temperature in the zone above the bed can easily fall below the value where ignition is possible. A common result is the emission of a yellowish-brown smoke which was for many years a feature of our urban landscape.

The combustion of coal (and other hydrocarbons) may lead to the emission of other unattractive substances too, including the oxides of sulphur and nitrogen. These, however, have only excited interest very recently, whereas smoke, because it is visible, has attracted adverse comments from the earliest times.

## Down-draught combustion

In 1680, Dalesme suggested a means of controlling the emission of smoke (Wright, 1964); in essence, he turned the simple coal fire upside down with a device illustrated in Fig. 2. A small fire is lit at the base of a refractory bowl, which is then filled with coal. Air enters at the top of the bowl, and combustion products leave at the bottom. This is generally called a 'down-draught' system, but in some applications this can be a misnomer, since the important technical feature is that the flame front now travels in the opposite direction to the primary air. It was reasoned that, in such a system, volatile matter evolved ahead of the flame front would be swept back by the air stream through the flame and into the incandescent part of the bed, where, to use a phrase of the time, it would be 'utterly consumed'.

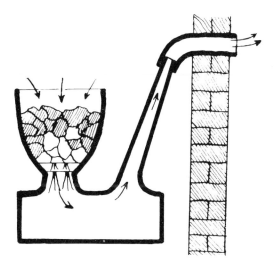

*Fig. 2. Dalesme's Heating Machine, 1680 (reproduced with permission from Wright, 1964).*

More recent studies have shown that volatile matter will only burn in these circumstances provided that it is well mixed with oxygen, is maintained at temperatures above 600°C and has sufficient residence time—about half a second—for the oxidation to go to completion. What is needed, in short, is sufficient turbulence, temperature and time.

Dalesme's principle was re-discovered in 1785 by James Watt, who patented it. Neither Watt nor his predecessor succeeded in making a practical version of their concept, however, probably because they did not have the necessary refractory materials, such as alloys or ceramics, at their disposal. Successful exploitation of the principle was made in the present century in the form of a number of mechanical stokers used in industrial steam raising.

## Industrial combustion — mechanical stokers

The underfeed stoker has been widely employed to fire small industrial boilers with outputs of about 0·5-3·0 MW. The principle of operation of the stoker is shown in Fig. 3. Coal is fed from the bottom left by a worm feeder into the bottom of a retort, the vertical walls of which are contained within a plenum chamber to which primary air is delivered by a fan. The coal rises vertically in the retort which the air enters through tuyeres in its sides. The fire is ignited at the top and the flame front tends to move downwards, its speed being matched by the rising flow of coal. The system thus fulfils the requirement of the flame front travelling in the opposite direction to the primary air. Volatile matter from the coal mixes with the air and ignites as it passes through the incandescent top layer of the bed, and so smoke emission can be effectively controlled.

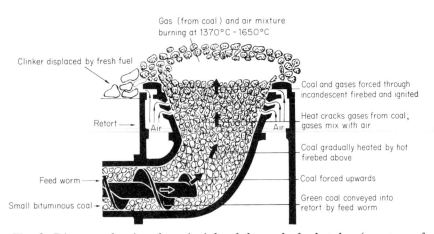

*Fig. 3. Diagram showing the principle of the underfeed stoker (courtesy of Ashwell Scott Ltd.).*

The chain-grate stoker is used in larger shell and water-tube boilers; an example is shown in Fig. 4. Coal is fed from hoppers by gravity to a grate which consists of an endless chain extending into the boiler. The horizontal movement of the grate carries the coal into the combustion chamber. A thin layer of coal is carried forward on the top surface of the grate, while air is delivered beneath it. As the coal enters the combustion chamber, its top

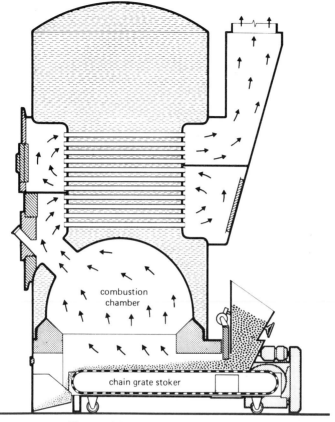

*Fig. 4. Chain grate stoker boiler.*

surface is ignited by radiation from a hot refractory arch. The flame front then travels down through the coal bed while the air comes up through it. When the coal reaches the end of the grate, the thin layer has been burnt out, and the residual ash is dropped into a container as the grate turns for the return journey.

## Pulverized fuel firing

All the combustion systems so far discussed have been fixed bed systems, in which coal in the form of particles sized, say, from 25 mm downwards, forms a static bed with gases passing through the interstices.

Extremely large coal-fired boilers, such as those used in power stations use entrainment systems in which the coal is carried into the boiler on a stream of the combustion air. This is called pulverized-fuel or p-f firing. The coal is first milled to very fine sizes (80% of particles are less than 200 μm), and is then entrained in an air-stream and delivered to a burner rather like that used for oil; two arrangements are shown in Fig. 5. Because of the very small particle size, the interior of a particle is heated very rapidly as it enters the combustion chamber, where flame temperatures of 1400°C are common, and as a result any volatile matter is quickly released. In the intensely turbulent conditions of the flame, the volatiles mix with oxygen, ignite and burn rapidly. The

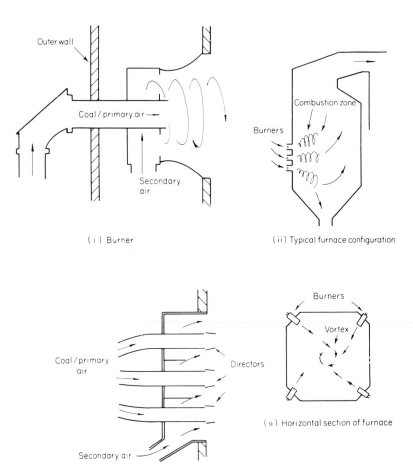

*Fig. 5. Pulverized fuel combustion of low and medium-rank coals.*
*Upper: Wall-fired boilers using turbulent flow burners. Primary and second-ary streams admitted with some swirl through concentric annular openings. Lower: Corner-fired boilers. Primary and secondary streams enter through alternate ports arranged vertically at the corners of the furnace, directed so as to generate a vortex in the combustion chamber.*

principle difference between this and oil combustion is that the residual particles of carbon have a longer burn-out time than oil droplets, and so the coal flame is longer. On the other hand, the presence of solid particles, coal and ash, gives the flame a high emissivity. Furnaces burning up to 220 tonnes of coal per hour are used in power stations producing up to 660 MW. Over half our production of coal in the UK is burnt each year by this method, with combustion efficiencies of over 98%.

One drawback of the method, however, is that the ash particles in the coal are raised to temperatures near 1400°C, and some mineral constituents will soften and glaze at these temperatures, while others will volatilize. If the ash particles are still soft when they enter the convective heat transfer part of the boiler, there is a possibility that they will form gluey deposits on the cooling tubes. Corrosive effects may also become apparent at these high temperatures.

## Fluidized bed combustion

The methods described above are well established. The next sections of this chapter deal with a new method of combustion which in some forms is just reaching the stage of industrial application, and in others is still under development. The method, called fluidized bed combustion, is of interest because it offers the possibility of several advantages, including the reduction of boiler costs, the capacity to burn a wide variety of fuels and better control of pollution.

The principle of the fluidized bed is illustrated in Fig. 6. Consider a cylinder containing a bed of inert solid particles, such as sand, supported on a grid such as fine wire mesh or a punched plate. Sand cannot fall downward through the grid, but a pumped supply of gas, such as air, can pass upward through it.

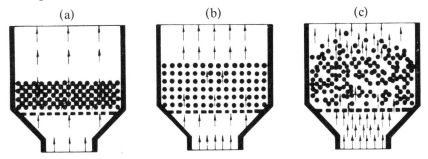

*Fig. 6. Principle of the fluidized bed:*
*(a) Gas velocity less than the fluidizing velocity,*
*(b) gas velocity approximately equal to minimum fluidizing velocity,*
*(c) Normal operating condition with gas velocity approximately three times minimum fluidizing velocity.*

If air is blown upward through the sand, and the gas velocity is increased, the air will at first pass between the sand particles (Fig. 6a), but as the velocity increases the individual particles will eventually be supported on the rising gas stream (Fig. 6b). The particles will begin to move about, and at still higher gas velocities some of the air will pass through the bed in bubbles. The sand bed takes on the appearance, and some of the properties, of a briskly boiling liquid, and we have a fluidized bed (Fig. 6c).

If the sand is now heated, and coal is injected continuously into the bed, the coal will burn very rapidly and the whole body of the sand will assume a uniform high temperature. This, in its simplest form, is a fluidized bed combustor.

There is nothing particularly new about it. Around the world a number of fluidized bed combustors are in use on a commercial scale, which use the Ignifluid system invented by Godel (1966). He recognized that a fluid bed offered higher heat release rates than conventional systems, and he devised a combustor in which the bed temperature is about 1300°C. The coal ash slags and forms lumps of clinker which fall to the bottom where they are removed by a moving grate. Hot gases leaving the surface of the bed are then passed through a conventional boiler.

Fluidized beds have another important property, however, in that heat transfer to cooling surfaces immersed within them is very high. The constant movement of the solid particles in the bed brings a hot particle into contact with a cooling surface and then removes it after a relatively short time, replacing it with another hot particle. The kind of fluidized combustor described below is one which operates at a lower temperature than the Ignifluid system, say at 800°C. At these low temperatures the ash does not soften and cooling surfaces can be immersed without fear of deposition upon them.

The potential advantages claimed for such a combustor, relative to a pulverized fuel furnace, are:

(1) High heat release rate—up to 3 MW $m^{-2}$ of bed area,

(2) Efficient heat transfer within the bed—less boiler surface required; potential for higher steam temperature,

(3) Temperature below 1000°C—reduced corrosion and fouling; low emission of oxides of nitrogen; retention of sulphur dioxide by limestone; non-sintered ash.

I hope to indicate in this chapter how far those advantages have so far been justified in the very active research and development programme which is going on all over the world.

Figure 7 shows a schematic fluidized bed boiler working at ambient pressure. This consists of a plenum chamber delivering the air for fluidization and combustion and a bed of inert solids (coal ash is a very convenient medium). Submerged in the bed are heat exchange tubes, and below them, fuel injection points. Above the bed is a freeboard space to allow for the violent splashing of particles at the bed surface, a convection pass and means for collecting and refiring elutriated dust. A boiler could consist of several beds in parallel, perhaps doing different duties.

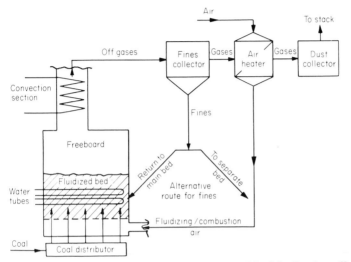

*Fig. 7. Diagrammatic arrangement of fluidized bed boiler installation.*

## III. DESIGN FACTORS IN FLUIDIZED BED COMBUSTORS

### Particle size and air velocity

The rate of combustion is determined by the air supply, and hence by the fluidizing velocity and operating pressure. Velocities from about $0.4$ m s$^{-1}$ to about 4 m s$^{-1}$, measured at bed temperature, have been used, giving corresponding heat release rates from $0.3$ to $3.0$ MW per square metre of bed area at atmospheric pressure. The size range of the bed material must be such that it is well fluidized at the chosen air velocity, and one way of ensuring this is to crush the feed coal to a size where it will produce ash in the desired size range. For example, coal sized $1.5$ mm-0 mm is appropriate for a velocity of $0.6$ m s$^{-1}$, while 6 mm-0 mm is appropriate to 4 m s$^{-1}$. Provided the coal is distributed uniformly across the bed, the volatiles and carbon monoxide produced are mostly burned within the bed, and the combustion of the small proportion which escapes is completed within about $0.3$ m of the bed surface.

An important feature of a fluidized bed combustor is that only a small concentration of coal, <5%, is necessary to sustain combustion and at typical operating conditions the concentration is less than 1% (Dainton and Finlayson, 1977).

The coal feed will, of course, contain a proportion of fine material, and inevitably some of these fines will be elutriated before they can be completely burned. This material may continue to burn in the freeboard space, but unburnt material leaving the furnace is readily collected by cyclones and, if necessary, can be burned by re-firing to the original bed or a separate one.

## Heat transfer

In addition to radiation, there is excellent convective heat transfer to surfaces immersed in the bed. Figure 8 shows the heat transfer coefficient—bed to tube—as a function of bed particle size. It will be seen that the coefficient decreases as the particle size increases. This forces the designer to compromise. If he uses a higher fluidizing velocity, his combustion intensity will be higher and his boiler smaller, but a higher velocity requires the use of larger particles, and the heat transfer coefficient drops. For atmospheric pressure operation a useful compromise is to use coal crushed to 3 mm top size, and a fluidizing velocity of about 2·5 m s$^{-1}$.

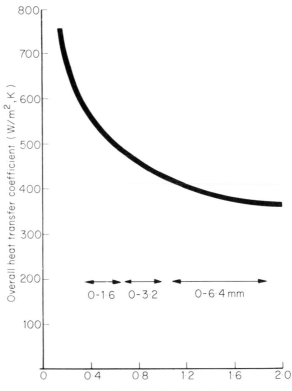

*Fig. 8. Heat transfer coefficient from a fluidized bed to an immersed tube, as a function of bed particle size.*

The major disadvantage of the system is the pressure drop across the distributor and bed itself, which necessitates the use of significant fan power, and this calls for another compromise. The volume of the bed is determined by the need to submerge the tubes. If we make the bed shallower in order to decrease the pressure drop, it will also have to be wider and the capital cost will be greater. A typical compromise here is a bed depth of about 1 m.

### Freeboard

The violent bursting of gas bubbles at the bed surface tends to fling solid particles high into the freeboard space above. This bubble action is important in the lateral transport of coal, but it is desirable that ejected particles should return to the bed. For this reason, a generous freeboard is often allowed and careful design of the convective tube bank in this space can help to deflect particles back into the bed.

### Temperature

A fluidized combustor works in a rather limited temperature range. At low temperatures of about 700°C some volatile matter and carbon monoxide escape without combustion, and efficiency falls. The upper limit is determined by the softening temperature of the coal ash which constitutes some or all of the inert solids in the bed. The softening temperature will vary from coal to coal, but for most British coals it is undesirable to raise the bed temperature above 1000°C, and it is usual to operate between 750 and 900°C.

With this low bed temperature, the temperature difference between the heat source and the working fluid is much less than in other systems, and it might be thought that heat transfer rates would be low. Happily, the heat transfer coefficient to submerged tubes is so high as to swamp the effect of the temperature difference.

### Coal injection

Volatiles and carbon monoxide are for the most part burned within the bed, provided that the coal is evenly distributed. It is the designer's job to ensure that this condition is fulfilled, and it presents him with a problem. In a highly agitated bed, some of the gas passes upward in bubbles which grow by coalescence as they rise, the velocity of the bubbles exceeding the mean gas velocity. These bubbles entrain particles in their wake, lifting them from the bottom of the bed. In their upward travel the bubbles push some coal particles sideways, and when they burst at the surface, particles are spread radially across the surface and then sink. The net effect of this action is that coal injected near the bottom of the bed is mixed rapidly in the vertical plane. Transport in the horizontal plane, however, is slower.

If, in a large bed, coal feed points are widely separated, then in their neighbourhood the coal concentration will be relatively high, while mid-way between the points it will be relatively low. The spacing of the coal feed points must be such as to ensure that the coal distribution is sufficiently uniform to prevent excessive loss of unburnt volatiles and carbon in the immediate vicinity of the coal feed points.

The necessary design data have been obtained in tests in which a sample of coal was irradiated so that one of the ash constituents (sodium) became radioactive. Small pulses of the 'labelled' coal were then injected into a large, cold fluidized bed. Samples of the bed solids were removed from other points in the bed over a period of time, and the concentration of coal determined by measuring the activity of the samples. In this way the lateral transport of the coal was measured and it was possible to estimate the concentration gradients within the bed of a large combustor.

Fluidized beds can also be used to burn gas or oil. The principal difference is that these two fuels simply pass vertically through the bed, while coal moves horizontally as well. For this reason fuel feed points for fluid fuels have to be much closer together than is the case with coal.

## Control and ignition

Some turn-down can be achieved by a reduction in fluidizing velocity, with a corresponding reduction in fuel feed rate. In this condition the bed operates at a lower temperature and a limit is reached when the combustion efficiency starts to fall.

Control over a wider range can be achieved by operating the bed as a number of separate units. The plenum chamber is divided into several sections, each separately supplied with air, and each corresponding part of the bed has a separate fuel supply. Turndown may be achieved by shutting off the air and fuel supply to one or more sections, causing it to defluidize or 'slump'. Because the thermal conductivity of the slumped ash is very low, this section will cool down only slowly, and it is possible to resume combustion merely by restarting the air and fuel supplies.

To light a bed from cold, a number of schemes have been tested, including burning gas within the bed and the use of submerged tunnel burners. Another method is to use a flame which is directed on to the surface; when the bed is fluidized, particles at the top are heated, and heat is transferred downward until the ignition temperature is reached.

In a multi-segment system it is possible to have one without cooling tubes. In this way a boiler could be kept in a 'hold-fire' condition for long periods at very low load.

## Control of sulphur emission

In the United States most of the interest in fluidized combustion centres around the possibilities it offers for pollution control, but at the present time these considerations are not foremost in the United Kingdom. In the UK there is a high stack policy which has ensured that, while the amount of sulphur dioxide emitted over the past decade has increased steadily, the ground level concentration of it has declined. However, we are a small island and recently our neighbours have been pointing out that if our sulphur dioxide is not falling on us, it may be falling on them. The case has yet to be proved, but there is a possibility that in the future our pollution control legislation may change towards the kind in force in the United States.

A fluidized bed is an excellent medium for contacting gases with solids, and this can be exploited in a combustor since sulphur dioxide emission can be reduced simply by adding limestone or dolomite to the bed. The sulphur oxides react to form calcium sulphate, which leaves the bed as a solid with the ash. The theoretical additive requirement is that limestone amounting to 3% of the coal feed should be added for each 1% sulphur in the fuel. In practice, under the most favourable operating conditions of low fluidizing velocity, efficient fines recycling and a temperature of between 800°C and 850°C, the theoretical addition retains about 80% of the sulphur, while double the theoretical rate retains about 95%. The retention efficiency falls at both higher and lower temperatures, at higher fluidizing velocities and inefficient refiring of unreacted fines. Some data are shown in Fig. 9. Dolomites require about twice as much weight added to obtain the same effect, since the magnesium component of dolomite does not react.

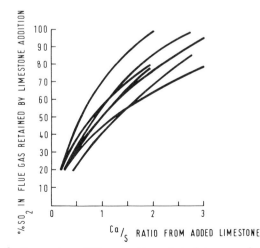

*Fig. 9. Percentage SO₂ retention plots for a number of coals.*

Limestones and dolomites from different sources vary in their effectiveness in reducing sulphur dioxide emission, because of differences in the pore structure produced on calcination.

## Oxides of nitrogen

The low combustion temperature eliminates the formation of nitrogen oxides from atmospheric nitrogen. The extent of nitrogen oxide formation from fuel nitrogen depends on the operating conditions, but emissions from fluidized combustors working at atmospheric pressure are lower than those found in conventional systems.

## Corrosion and erosion

Boiler surfaces may be subject to the corrosive effect of some constituents of the fuels used in them. For example, superheater tubes in large, pulverized-fuel boilers may be corroded by a process involving the alkali metals sodium and potassium, which occur in coal ash. Moreover, where high gas velocities occur in boilers, particles of hard ash may cause erosion of surfaces.

Because the temperature of fluidized bed combustors is very much lower than that of pulverized-fuel flames, it might be supposed that the quantity of alkali metals volatilized from the ash would be less, and the corrosive effects due to this would be reduced. Extensive tests in fluidized beds have shown that this is the case.

The tests have shown that for bed temperatures up to 850°C and metal temperatures up to 650°C, a range of alloys is available which will give satisfactory boiler tube performance. However, sulphidation does increase when limestone is added to the bed for the control of sulphur emission, and further work is under way to identify materials for use at metal temperatures above 650°C (Rogers *et al.*, 1977).

Although a fluidized bed is a turbulent assembly of solid particles, erosion of submerged tubes is not a problem. Since the temperature is low, the ash particles are soft and unglazed, and the average particle velocity is only approximately 1 m s$^{-1}$.

## Coal characteristics

Because the combustion temperature is low, the ash fusion characteristics of coals present problems only in exceptional circumstances. Since the coal concentration is so low in the bed, one could burn coals with ash contents up to, say, 70%. Variation in the ash content of the feed coal is no embarrassment, provided that the coal feed rate can be varied sufficiently to cope with the changes in calorific value. A wide range of caking powers and volatile contents is acceptable—coals ranging from very strongly caking coal to non-caking coals of high volatile content. The exception, however, is anthracite; because of its very low reactivity, a higher proportion of fine carbon is elutriated before it can be burnt.

# IV. FLUIDIZED BED COMBUSTION AT ELEVATED PRESSURE

So far the discussion has considered operation at ambient pressure, but as far as power generation is concerned, fluidized combustion becomes even more attractive a proposition at high pressure. The pressure energy of the off-gas must be used by expanding it through a gas turbine which drives the air compressor and also generates power. The combination of steam turbine, driven by steam raised in tubes submerged in the bed, and gas turbine driven by the off-gas, gives rise to an increased efficiency of power generation.

At constant fluidizing velocity, high pressure operation increases the rate of oxygen supply and thus allows the combustion rate to be increased in proportion to the pressure. The area of heat transfer surface must also be increased in proportion to pressure, and the bed is consequently deeper than in combustion at atmospheric pressure in order to submerge the extra tubes. Pressure has no significant effects on heat transfer characteristics in the bed, or on combustion efficiency except in so far as the deeper bed is an advantage.

Figure 10 shows a section of the pressurized fluidized bed rig at Leatherhead which has been used for an extensive research programme. The bed is contained in a 2 m diameter pressure shell and burns up to 500 kg h$^{-1}$ of coal at pressures up to 6 bar. The off-gases are cleaned in cyclones before passing at a velocity of 125 m s$^{-1}$ over a cascade of static turbine blades to assess the extent of erosion, corrosion or deposition. The dust burden of the gas is generally about 200 ppm and most of the dust is less than 20 $\mu$m in size. It is flaky and friable, and shows no sign of fusion.

At bed temperatures of about 800°C there has been no evidence of erosion or corrosion of the blades, which are lightly coated with a layer of ash which reaches an equilibrium thickness in about 20 hours. This layer can be removed by conventional 'on-line' cleaning methods. It therefore seems feasible that gas turbines can be operated by pressurized fluidized bed combustors.

Sulphur retention may also be achieved in an elevated pressure system, where it is preferable to add dolomite rather than limestone (Roberts et al., 1975). Pressurized systems have an added advantage in that emissions of nitrogen oxides are significantly lower than at atmospheric pressure (Shaw, 1976).

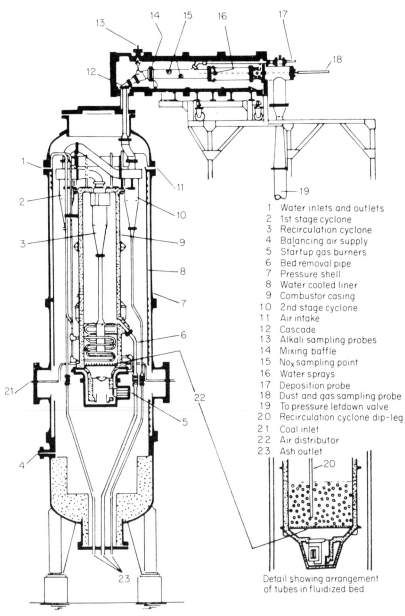

1   Water inlets and outlets
2   1st stage cyclone
3   Recirculation cyclone
4   Balancing air supply
5   Startup gas burners
6   Bed removal pipe
7   Pressure shell
8   Water cooled liner
9   Combustor casing
10  2nd stage cyclone
11  Air intake
12  Cascade
13  Alkali sampling probes
14  Mixing baffle
15  No$_x$ sampling point
16  Water sprays
17  Deposition probe
18  Dust and gas sampling probe
19  To pressure letdown valve
20  Recirculation cyclone dip-leg
21  Coal inlet
22  Air distributor
23  Ash outlet

Detail showing arrangement
of tubes in fluidized bed

*Fig. 10. Pressurized fluidized bed rig.*

## V. COSTS OF FLUIDIZED COMBUSTION

In a number of design studies, the costs of fluidized combustion have been compared with those of pulverized fuel boilers for large power stations. In Table I, results of a recent US study are presented. On the basis of these figures, the advantage would seem to lie with atmospheric pressure fluidized combustion if interest rates are high and fuel costs low. In circumstances where interest rates are low and fuel costs high, then a pressurized system would appear to be preferable.

*Table I. Comparison of power station systems.*

|  |  | Capital cost $ per kW | Efficiency % |
|---|---|---|---|
| Pulverised fuel | (A) | 620 | 36·2 |
| Pulverised fuel | (B) | 835 | 31·8 |
| Atmospheric f.b.c. | (C) | 632 | 35·8 |
| Pressurized f.b.c. | (C) | 723 | 39·2 |

(A) Without sulphur dioxide retention
(B) Sulphur dioxide retention by flue-gas washing
(C) Sulphur dioxide retention by use of limestone/dolomite in the fluidized bed.

*Source* Meeting on Phase II of ECAS project: General Electric designs in Report No. NASA CR-134949, Volume II, part 2, December 1976.

## VI. RECENT DEVELOPMENTS AND APPLICATIONS

### Power generation

Interest in fluidized bed combustion has now reached the point where a number of demonstration and prototype plants have entered service or are nearing completion. In the field of power generation, a 30 MW demonstration plant working at atmospheric pressure has been built and operated by the Foster-Wheeler Corporation at Rivesville, West Virginia. Under the auspices of the International Energy Agency, a large, experimental pressurized combustion rig is under construction in the UK. This will have a bed area of 4 m² and be able to operate at temperatures up to 950°C and pressures up to 12 bar. The nominal maximum heat output will be 85 MW, over 50 times as great as that of the Leatherhead rig.

### Waste disposal

The disposal of 'tailings', the fine, wet waste material from coal washeries, presents a problem to the coal industry. This material is mostly water, but contains a proportion of very fine particles of mineral matter and coal. The use of large lagoons is environmentally undesirable, and disposal on waste tips requires great care.

The current technique employed for disposal is to thicken the tailings until the composition is roughly 50% water, 30% mineral matter and 20% coal. More water is then physically squeezed out in large filter presses, and the resulting filter-cake is carefully mixed with other wastes and disposed on a tip.

In the thickened tailings at about 50% moisture content, there is enough calorific value in the coal to drive off all the water and raise the mineral matter to a temperature approaching 1000°C. It is difficult to ignite and sustain combustion in a material which is 50% water and 30% ash, but it can be done in a fluidized bed combustor.

The European Coal and Steel Community is giving financial support to an investigation of this method, and a pilot plant burning 1 t $h^{-1}$ of thickened tailings is now in operation. The system is rather different from the boilers described hitherto, and Fig. 11 shows the form of combustion chamber, in which the wet slurry is sprayed down on the surface of the bed.

Variation in the calorific value of the input can be dealt with by the injection of small quantities of auxiliary fuel beneath the surface of the bed. The ash which is the principal product emerges up-graded in size from the original mineral matter, and an important part of the present investigation is to study the influence of the design and positioning of the slurry nozzle on the size distribution of the ash. The ash can be safely disposed of, but beyond that there are possibilities of heat recovery from the system and commercial applications for the ash as an aggregate.

## Industrial drying and steam-raising

In the 1960s oil became cheaper than coal in many parts of Great Britain, particularly on the industrial market. Traditionally, this market was dominated by customers who used horizontal shell boilers, generally burning washed, graded coals about 25 mm to 12 mm in size. When they realized that oil was not only more conveniently handled, but cheaper, they naturally converted to oil. Moreover, throughout the 1960s many of the old boilers were replaced by compact, packaged boilers, purpose-designed for oil.

Over the last four years, however, because of the rapidly changing situation in fuel prices, many customers have expressed a strong interest in converting back to coal, but this is not easy. The combustion intensity which can be achieved with oil is about twice that which can be achieved with coal fired by chain-grate stokers. Consequently, an oil-fired boiler of a given output is smaller and about half the capital cost of a coal-fired boiler.

In order to get coal back into this market, it is necessary to devise new ways of firing coals with much greater intensity than conventional stokers can

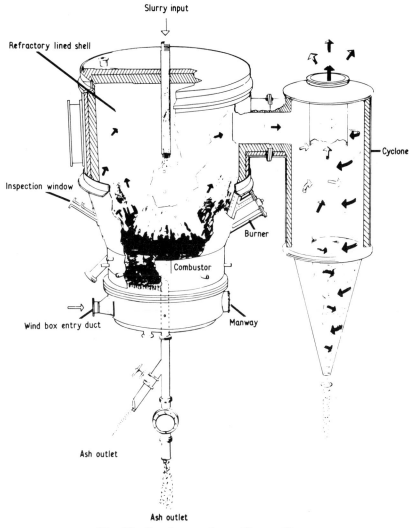

*Fig. 11. Combustor for colliery tailings.*

achieve. The NCB is looking at methods of doing this, and one possibility is
the use of fluidized beds. The fluidized bed offers a means of obtaining the
required burning rate, but in this application the geometry is at first sight
unfavourable. The systems described above tend towards a small, cross-
sectional area and a generous freeboard above. The geometry of a shell boiler
is completely the reverse, since its combustion chamber is a long horizontal
tube usually about 1 m in diameter and never more than 1·5 m.

Two lines of approach are being followed. First, the development of compact vertical shell boilers fired by fluidized beds and second, the conversion of existing horizontal shell boilers to fluidized firing. This has led to the use of much shallower beds than were formerly considered acceptable, and also to the discovery that when washed coals of low ash content are fed into these boilers, the top size of coal can be raised to 25 mm or even 50 mm. As a result of using larger coal sizes, the proportion of fuel which is burnt before the residual particle is elutriated is greatly increased. A number of prototype boilers are now under test in field conditions.

In the area of industrial water-tube boilers, Babcock and Wilcox have converted one of their boilers in Scotland to fluidized bed firing, using a bed area of 9·3 m², and giving an output of 20 000 kg h⁻¹ of steam. A section of this boiler is shown in Fig. 12. It has been operated successfully for more than a year, using a wide variety of fuels, and the company is now offering fluidized bed boilers to the commercial market.

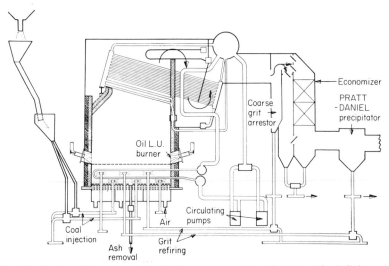

*Fig. 12. Industrial water-tube boiler converted to fluidized bed firing.*

Apart from boilers, fluidized bed combustors have applications for the production of hot gas. Figure 13 shows a schematic drawing of a kiln used to dry grass for the preparation of cattle food. These kilns are commonly fired by oil, but recently three have been converted to fluidized bed firing of coal. There are no tubes in the bed; temperature is controlled by the excess air used in the system (i.e. air above that quantity required for stochiometric combustion) and the hot off-gases are fed to the kiln. The bed has several

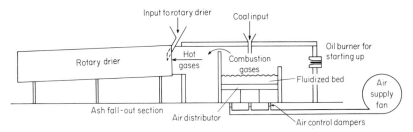

*Fig. 13. Fluidized bed heated kiln for grass drying.*

compartments which can be run independently under conditions of varying load, and the combustor is quite simple to operate. Coal-fired grass driers are now being offered on the market.

## VII. DOMESTIC CENTRAL HEATING

The final part of this chapter is concerned with one aspect of the very simplest combustion system such as may be used in domestic heating.

The British people still have a very marked addiction to the open fire, and over 200 000 new appliances of this type are sold every year. The simplest open fire works on the up-draught principle, illustrated in Fig. 1, and it will be remembered that one of the main disadvantages of such an appliance is the emission of smoke. One way of overcoming this problem is to manufacture a suitable smokeless fuel, but such a process is a rather inefficient way of using the energy available in coal.

How much better it would be if the fire could itself consume the smoke! The suggestion of Dalesme in 1680 was that if the flow of air could be reversed, the problem could be solved. Figure 14 shows a cross-section of a modern fireplace in which these objectives have been realized.

The bed of coal is supported by a grate and the fire is ignited by a chemical lighter placed at the bottom. Primary air passes through the front aperture and down through the bed, in the opposite direction to the advancing flame front. Volatile matter is thus heated and ignited as it passes through the throat at the back on its way to the flue. To make sure there is efficient combustion, secondary air passes down a duct at the rear of the fire, being heated as it goes, and mixes with the volatile matter in the throat. The lower part of the fire is surrounded by insulating refractory to make sure that the temperatures remain high enough, while in the upper part there is a boiler through which the product gases pass on their way to the chimney. This boiler supplies domestic hot water, and heats up to six radiators. The appliance can burn a reasonably wide range of high-volatile coals with a high efficiency and reduces smoke emission to the levels required by the Clean Air Act.

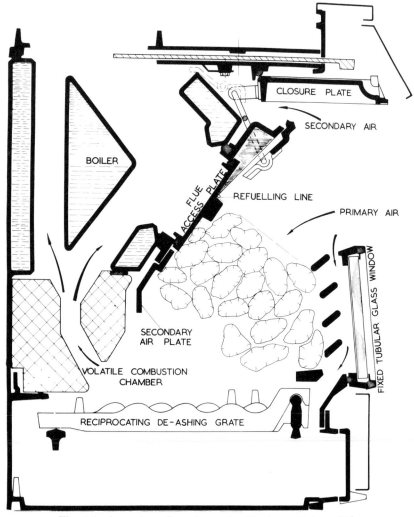

*Fig. 14. A modern fire-place: the Rayburn Prince '76.*

## VIII. CONCLUSION

In Chapter 8, techniques for the preparation of liquid fuels from coal are discussed, and undoubtedly, as oil supplies run short towards the end of this century, these methods will assume greater industrial importance. However, in the medium term, the cheapest way of making oil from coal, so to speak, is to burn coal for steam-raising where oil was burned before, thus releasing liquid fuels for use in more specialized markets, and it is this that will make coal combustion so important in the years immediately ahead.

# REFERENCES

Dainton, A. D. and Finlayson, P. C. (1977). 10th World Energy Conference, Istanbul, paper.

Godel, A. A. (1966). *Rev. Gen. Thermique* **52,** 349.

Rogers, E. A., Page, A. J. and La Nauze, R. D. (1977).

Eurocor, 6th European Congress on Corrosion, London.

Roberts, A. F., Stantan, J. E., Wilkins, D. M., Beacham, B. and Hoy, H. R. (1975). Institute of Fuel Symposium on Fluidized Combustion, London. Paper D.4.

Shaw, J. T. (1976) 2nd International Conference on the Control of Gaseous Sulphur and Nitrogen Compound Emission, University of Salford.

Wright, L. (1964). "Home fires burning." Routledge and Kegan Paul, London.

# 7 Gasification of Coal

## A. D. DAINTON

*Director, Coal Research Establishment, National Coal Board*

## I. INTRODUCTION

The manufacture of an inflammable gas from coal dates from the work of William Murdoch at the end of the eighteenth century, and for over 150 years gas-works were a common sight all over Western Europe. The product, called 'town's gas', had a calorific value of about 17 MJ m$^{-3}$ and was made by the thermal decompostion of coal in a closed retort. The gas was a mixture of hydrogen, methane and carbon monoxide, and the process produced large quantities of residual carbon, called gas coke, which was also used as a fuel. In the late 1950s, this process was displaced in the UK by the production of gas from oil-based feedstocks, which, in turn, gave way to natural gas from the North Sea.

However, before the demise of the gas-works, attention was being paid to new ways of gasifying coal. Instead of deriving the gas from a small proportion of the feedstock, the new aim was to gasify the whole, or virtually the whole of the coal, a method called 'total gasification'.

Although natural gas has not been used for very long, the end of the supply is already in sight and there is a renewed interest throughout the world in coal gasification. This new interest centres on total gasification, and it is the aim of this chapter to describe, in broad terms, a number of different ways in which coal may be gasified, and to discuss applications of gasification to power generation, chemical synthesis and the production of substitute natural gas.

## II. REACTIONS IN A GASIFIER

### Chemical

British coals contain between 77 and 94% carbon, and it is a convenient simplification to think of coal gasification mainly in terms of reactions between carbon and suitable gases. The main reactions which take place in a gasification system are listed below:

| | | |
|---|---|---|
| Combustion | $C + \frac{1}{2}O_2 \longrightarrow CO$ | Heat released |
| | $CO + \frac{1}{2}O_2 \longrightarrow CO_2$ | Heat released |
| Gasification | $C + H_2O \longrightarrow CO + H_2$ | Heat absorbed |
| | $CO_2 + C \longrightarrow 2CO$ | Heat absorbed |
| Hydrogenation | $C + 2H_2 \longrightarrow CH_4$ | Heat released |
| Shift Reaction | $CO + H_2O \longrightarrow CO_2 + H_2$ | Heat released |
| Methanation | $CO + 3H_2 \longrightarrow CH_4 + H_2O$ | Heat released |
| Pyrolysis | $4\,C_nH_m \longrightarrow mCH_4 + (4n-m)C$ | Heat released |

The key reactions are those between carbon and steam, and carbon and carbon dioxide. These reactions are endothermic, and the main reason for encouraging combustion of carbon in a gasifier is to generate heat to drive the desired endothermic gasification reactions. Hydrogenation permits the direct production of methane from carbon, but a more usual route is to use the shift reaction to increase the proportion of hydrogen in the synthesis gas, and then proceed with methanation, a strongly exothermic reaction promoted by catalysts.

The last reaction on the list is pyrolysis, the thermal decomposition of hydrocarbons in the coal. This produces a range of products depending on the temperature and rate of heating, and is the basis of the old town's gas process. Some reactions of this kind will take place in any gasifier.

### Supplying heat

The necessary heat for the carbon-steam and carbon-carbon dioxide reactions is commonly generated by burning part of the carbon in the gasifier. If this is achieved by supplying pure oxygen, the product will be a synthesis gas, mostly hydrogen and carbon monoxide, with a calorific value of 10 MJ m$^{-3}$, which can be used either as a fuel or for chemical synthesis, including methanation. If combustion is achieved by supplying air, which is of course much cheaper, then the product gas will be heavily diluted with nitrogen, and have a calorific value of 3 to 4 MJ m$^{-3}$. This low calorific-value (CV) gas will be unsuitable for long-range distribution or chemical synthesis.

A possible alternative is to supply heat indirectly to the gasifier from an external source, such as a separate combustor or a nuclear reactor. This presents problems of heat exchange, but the product gas has a high calorific value and is directly suitable for further synthesis.

## III. GASIFICATION METHODS

### General

The main technical difficulties in gasification do not arise from the chemistry of the process, but from the nature of coal. In principle, any process must achieve efficient contact between the solid carbon surface and the reacting gases, and this will be affected by the size distribution of the coal particles in the system. Moreover, most bituminous coals soften when heated to temperatures between 400 and 500°C, and particles in this condition may coalesce to form larger solid structures (see Chapter 3). Efficient gas-solids contacting may therefore require careful size preparation of the feedstock, the selection of non-caking coals, or the pre-treatment of the coal in order to destroy its caking capacity.

The coal will also contain mineral matter, which can be withdrawn from the gasifier as a dry (unfused) ash, as clinker or as molten slag, depending on the mineral composition and the temperature reached in the process. It is clearly desirable that this ash should be removed with the minimum content of carbon, in order to maximize the conversion efficiency.

A third consideration is efficient heat exchange between the exothermic and endothermic zones in the gasifier.

Attempts at solving these problems have led to the development of a number of different methods of gasification. The more important fall into the same classifications as the combustors discussed in Chapter 6, namely those in which the coal particles form a fixed bed, are entrained in the gas stream or form a fluidized bed. Examples of several types are described below.

### Fixed-bed gasification: the Lurgi gasifier

The Lurgi gasifier is shown in Fig. 1; it is a cylindrical, water-cooled vessel operated at about 20 bar. The bed of coal particles is supported on a massive rotating grate, and oxygen and steam are injected from beneath. The maximum temperature is not allowed to exceed about 1000°C so that the ash does not slag. To ensure a steady downward fuel flow with uniform distribution of reactants and product gas, the coal is prepared in a range of sizes generally not exceeding 6:1, with fines removed. The stable, slow-moving fuel bed is replenished in small increments via an upper lock-hopper, through a revolving coal distributor, while ash is removed by rotation of the grate.

In its downward passage through the vessel, the coal passes through a number of well-defined zones. The top of the bed may be termed the pre-heating zone where the coal is quickly heated to 400°C. It then enters a degasification zone in which methane, higher hydrocarbons and tars are evolved. Subsequently, in the gasification zone, carbon monoxide and hydrogen are produced. The combustion zone and the ash layer lie just above the grate.

The gas leaves at relatively low temperature, between 450 and 600°C, and is passed to a scrubber where tars and dust are removed and the temperature is reduced to 185°C.

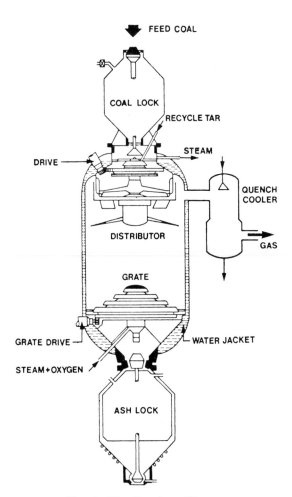

*Fig. 1. The Lurgi gasifier.*

Lurgi gasifiers have been in use since 1936. Sixty-five commercial units have been built, mostly oxygen blown, but some using air to produce low CV gas. The coals used have been mostly lignites, anthracite and weakly caking bituminous coals. This long and proven employment is the important advantage of the Lurgi process. The disadvantages may be said to be:

(1) mechanical complexity and consequent high costs;
(2) small capacity;
(3) production of tars which have to be removed;
(4) high steam requirements (to keep temperature low);
(5) selectivity in the size and caking capacity of coals.

## Entrainment gasification—the Koppers-Totzek (KT) process

In the Koppers-Totzek process shown in Fig. 2, the fuel is ground to very fine sizes (~200 μm) and injected in a stream of oxygen and steam into a squat, circular vessel which operates at approximately atmospheric pressure.

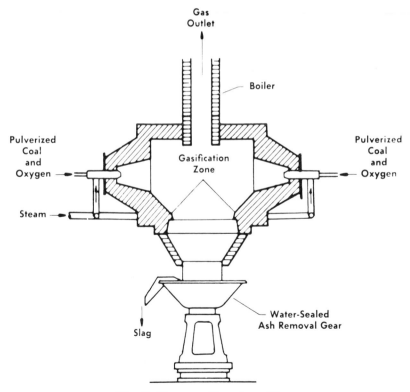

*Fig. 2. Koppers-Totzek gasifier.*

The reaction takes place at a maximum temperature above 1500°C, the reaction time being about 1 s. The carbon is consumed almost completely, and tars are gasified along with the carbon. The product gases leave the vessel at about 1500°C and rise into a waste heat boiler where they are cooled to 300°C, raising the necessary steam for the process. At the very high temperatures existing in the gasifier, the ash in coal becomes molten and more than half of it is removed at the bottom as a slag.

The high temperature results in the production of relatively high proportions of carbon monoxide and low proportions of carbon dioxide and hydrocarbons. The efficiency of steam conversion is satisfactory, but oxygen consumption is higher than in the Lurgi process.

The KT process has operated commercially since 1952, and 53 units have been built. It can handle a very wide variety of feedstocks, including oils, peat, lignite and strongly caking coals.

Its advantages can be summarized as:
    (1) ability to handle wide variety of coals;
    (2) no tars in the gaseous product;
    (3) gasifier is simple and easy to maintain;
    (4) capability of raising large amounts of process steam.
The disadvantages are:
    (1) has not yet been adapted to operate at high pressure;
    (2) air blown operation may not be feasible;
    (3) coal must be pulverized;
    (4) associated plant is large;
    (5) thermal economy is poor.

## Fluidized bed gasification — the Winkler process

One form of the Winkler gasifier is shown in Fig. 3. Fuel is injected via a worm into the lower part of the gasifier vessel where it is fluidized in oxygen and steam at about atmospheric pressure. The temperature of the bed is controlled at a relatively low level (800°C in this example). Particles thrown into the free-board space react at a higher temperature with secondary oxygen injected above the bed, and with steam escaping from the bed. Fine material carried out of the vessel passes into a cyclone, where much of it is collected and returned to the reactor.

The fluidized bed is attractive because if offers the possibility of efficient heat transfer from exothermic to endothermic regions. On the other hand, it is generally limited to temperatures at which the ash does not sinter, and gasification rates are low at these lower temperatures.

Winkler gasifiers came into commercial use in the late 1920s, and 36 units have been built. They have been most successfully employed with lignites or lignite cokes, which are very reactive and will gasify at the lower temperatures.

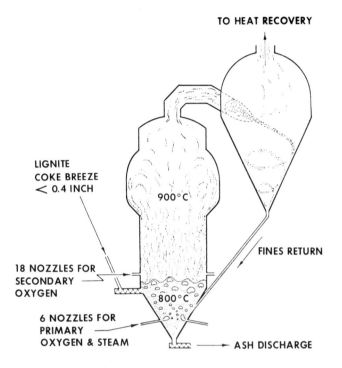

*Fig. 3. A late-model Winkler gasifier.*

The advantages of the Winkler process lie in the good temperature control and relatively easy solids handling. It can also be air blown if required.

The disadvantages are:

(1) The range of operating temperature is limited at the upper end by the requirement to avoid ash sintering, and at the lower end by the desire to avoid tar formation.

(2) Carbon loss tends to be higher than in other systems, since there are substantial losses on ash extraction from the lowest point, and also in the elutriated fines.

(3) Caking coals cannot be handled directly, i.e. they require some pre-treatment to make them non-caking.

## Comparison of the main processes

The operating conditions used in the three main steam/oxygen gasification processes are compared in Table I. Table II shows the composition of the primary products and their calorific value.

Table I. Operating conditions of the main steam-oxygen processes.

|  |  | Winkler | Lurgi | KT |
|---|---|---|---|---|
| Maximum gas exit temperature | °C | 980 | 540 | 1400 |
| Operating pressure | bar | 1 | 20 | 1 |
| Oxygen used, kg/1000 MJ gas |  | 34 | 17 | 47 |
| Ratio steam/oxygen |  | 3·5:1 | 4·5:1 | 0·6:1 |

These figures illustrate that:
(1) carbon dioxide decreases, and carbon monoxide increases with increasing temperature;
(2) methane increases with increasing pressure and decreasing temperature;
(3) the data on the Lurgi process suggest that gas composition does not vary appreciably with coal type.

Table II. Composition of primary products from steam-oxygen processes.

|  | Winkler | Lurgi | | KT |
|---|---|---|---|---|
| Fuel | Lignite | Lignite | Bituminous |  |
| $CO_2$ | 16·0 | 31·1 | 31·9 | 11·9 |
| CO | 45·1 | 15·3 | 17·8 | 55·6 |
| $H_2$ | 36·5 | 40·8 | 38·6 | 31·0 |
| $CH_4$ | 1·6 | 11·1 | 9·3 | 0·1 |
| Ethane and heavier | — | 0·7 | 1·1 | — |
| Nitrogen and argon | 0·8 | 1·0 | 1·3 | 1·4 |
| Total | 100·0 | 100·0 | 100·0 | 100·0 |
| CV (gross) MJ m⁻³ | 10·2 | 11·6 | 11·4 | 10·5 |

## Molten bath gasification

Apart from the three main classes of gasifier, some quite different types have been suggested in recent years. For example, one of the molten bath processes, the Atgas process, represents a radical departure (La Rosa and McGarvey, 1975). The objective is to produce a sulphur-free fuel gas and the process is based on the well-known affinity between iron and sulphur. Coal is dissolved in molten iron so that the volatile matter is pyrolysed and cracked to form combustible gases. The residual carbon and the sulphur dissolve in the iron and the carbon is then gasified by air blown through the bath

forming carbon monoxide. The sulphur diffuses to a lime-bearing slag
floating on the surface from whence it is removed, more limestone being
added to make up the slag.

## The COGAS process

The COGAS process employs a fluidized bed gasifier, but it differs from the
others in that it uses no oxygen, either pure or in the form of air, in the
gasification stage. The process has its origins in an earlier development
called COED, the Coal-Oil-Energy-Development. As originally conceived,
the COED process involved a mild, multi-stage pyrolysis of coal, generating
some gas and also tar which underwent hydro-treatment to produce a
synthetic crude oil. The residual char, which constituted about 75% of the
feed coal, was intended to be burnt in a power station to generate
electricity—hence the name Coal-Oil-Energy.

The process was demonstrated successfully on a large, pilot scale by the
Food Machinery Corporation of the US, but further exploitation was halted
by changes in US anti-pollution legislation. Under the new legislation,
sulphur emission from power stations was controlled to a level which
excluded the use of a high proportion of eastern US coals. The disadvantage
of COED was that most of the sulphur in the feed coal emerged in the char,
which then became less desirable as a power station fuel.

Following this set-back, a new process called COGAS was devised, a block
diagram of which is shown in Fig. 4. The first pyrolysis stage is the same as
before, but most of the residual char instead of being burned is gasified with
steam. A portion of the char is burned in a combustor, which heats a solid
medium by which heat is transferred to the gasifier.

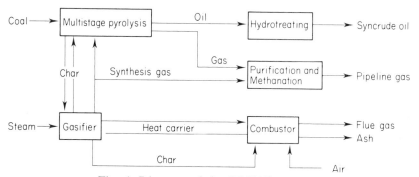

*Fig. 4. Diagram of the COGAS process.*

Most of the sulphur is converted to hydrogen sulphide in the gasifier and as such is easier to deal with than in the form of sulphur dioxide. The combustor and gasifier have been developed at the NCB's Leatherhead laboratory for the sponsors of the process, and the construction of a large demonstration plant in the US is now being considered.

## IV. ELECTRICITY GENERATION

### General

An important application of gasification in the future will be in the generation of electric power. In present-day practice, large quantities of coal are burned to raise steam in power stations, and electricity generation is coal's largest single market. In view of the wide-spread use of coal combustion, and the fact that combustion is generally simpler than gasification, it is necessary to explain why gasification should even be considered for this application. If coal is to be turned into a gas which is immediately burned to generate power, why not burn the coal directly?

There are two reasons for doing so, the first arising from a local concern for pollution. In the United States, for example, there are a number of power stations fired by natural gas, and their operators are already looking for alternative fuels as the supply of cheap methane declines. There are plenty of high-sulphur coals available, but the boilers are not suitable for direct coal firing. They can, however, be fired by a coal-derived gas, even if it is of relatively low calorific value, and the sulphur can be removed as hydrogen sulphide.

The second reason is more universal in that it concerns every nation which will be hungry for energy towards the end of this century. From the point of view of energy conservation, power production is a wasteful form of utilization, since the thermodynamic efficiency of generation is low, yet electricity is essential, at least for lighting and motors. In a situation where high cost fuel is in short supply it becomes increasingly desirable to improve the efficiency of electricity generation. The gasification route offers an opportunity for such an improvement.

### Steam cycles

In a conventional power station, fuel is burned in a combustion chamber and the hot gases are used to raise steam, which drives a steam turbine. Normally, steam turbines operate on the Rankine cycle, and in practice this has an efficiency of about 33%.

Over the years, various methods have been used to improve the efficiency, notably by taking steam from the first stage of the turbine and returning it to the furnace to be reheated before passing to the second stage. By this and other methods it is possible to raise the efficiency to about 38%, but it is thought unlikely that this figure will be improved any more, unless there is significant progress in materials of construction which will permit higher steam temperatures to be used reliably.

However, gas turbines work at higher inlet temperatures and the use of a gas turbine, in conjunction with steam turbines in a so-called combined cycle, presents the possibility of higher generating efficiency.

## Combined cycles

A gas turbine accepts gas at a high temperature ($\sim$1000°C) but also rejects it at a relatively high temperature, say, 600°C. If the reject heat is then used to generate more electricity in a steam cycle, we have a combined cycle. If the efficiencies of the individual cycles are 30%, then the overall efficiency will be $0.3 + 0.7 \times 0.3$, i.e. 51%.

The simplest kind of combined cycle is the waste-heat recovery type shown in Fig. 5. Waste heat from the gas turbine passes to a heat exchanger where steam is raised. With the presently available gas turbines, this configuration is not the most efficient one because the temperature at the outlet from the gas turbine is not high enough to allow an efficient steam cycle.

Figure 6 shows the exhaust-fired system in which supplementary fuel is burned in the steam-raising equipment to improve the efficiency of the steam cycle.

In the supercharged system shown in Fig. 7, the fuel is burned under pressure in the boiler system and then the exhaust gas passes through the gas

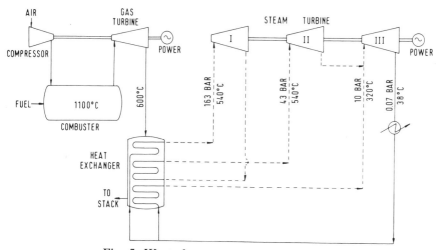

*Fig. 5. Waste-heat recovery combined cycle.*

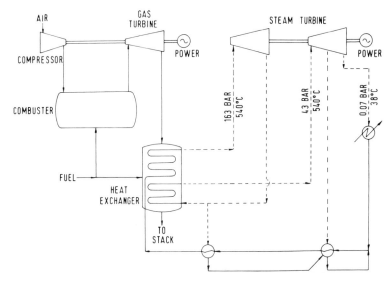

*Fig. 6. Exhaust-fired combined cycle.*

turbine. Here the reject heat from the gas turbine may be used to preheat the working fluid. This kind of cycle is that which was described in the context of pressurized fluidized combustion (Chapter 6).

Currently, industrial gas turbines have a limitation on the inlet gas temperature of about 1000°C. The use of pressurized fluidized combustion in a combined cycle giving efficiencies of 40% or more is therefore a real

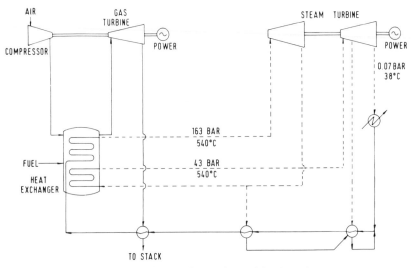

*Fig. 7. Supercharged combined cycle.*

possibility. However, the development of gas turbines is not at a standstill, and we may assume that within a decade or so industrial turbines will become available which have maximum inlet temperatures of 1200°C or even higher. There is no direct method by which pressurized fluidized combustion could exploit the potential improvements in efficiency which these new machines will offer, since the combustion temperatures would be so high that the ash would slag. However, if coal could be gasified, cleaned without cooling and the gas burned to raise its temperature to the maximum allowed by the gas turbine, then efficiencies of 45 to 50% are conceivable for inlet temperatures around 1200°C.

Work on gasification for power generation is further advanced in the US and Germany where, apart from other attractive features of the process, the somewhat greater ease of sulphur removal is an important consideration.

## Quality of gas for gas turbines

Provided that the combustion chamber and associated ducting are designed for the fuel which is to be supplied, its composition and calorific value are not significant for the gas turbine itself. Gas turbines have been operated satisfactorily on natural gas with a calorific value of 37 MJ m$^{-3}$ at one end of the scale and on blast-furnace gas with a calorific value of 3 MJ m$^{-3}$ at the other end of the scale. For economic reasons the tendency in current developments is to aim for low CV gas, using air and not oxygen as the heat-raising medium.

It is important that the gas should be clean enough to avoid fouling, corrosion and erosion. The degree of cleanliness will depend on the design of the turbine, on its operating conditions and on the operating life required. Generally speaking, the more extreme the conditions and the longer the turbine life required, the cleaner must be the gas. At present there is no clear-cut specification of the quality required for turbines fired by gas from the gasification of coal. Limits have been proposed by American gas turbine manufacturers (see, for instance, Foster, 1970, Fraley and Kumar, 1975, and Giramonti and Lessard, 1975). These may be summarized by saying that the concentration of particles larger than 20 $\mu$m (whether ash, carbon or tars) should be negligible and that the concentration of particles larger than 2 $\mu$m should be small—particles smaller than 2 $\mu$m are not regarded as erosive. There is no clearly defined limit for sulphur content.

The real problem arises with alkali metals. It is not known just how low the alkali metal content must be maintained in gases arising from coal gasification, but present indications are that the specification may have to be set so low as to create problems in meeting it. Fraley and Kumar (1975), for instance, say that the alkali metals and sulphur should be less than that required to form the equivalent of 5 ppm of alkali metal sulphate in the fuel when operating above 620°C.

# V. THE CURRENT STATUS OF WORK ON GASIFICATION FOR POWER GENERATION

The status of some of the development projects in hand will now be briefly reviewed, and again the topics can be classified under the general headings of fixed bed, fluidized bed, entrainment and molten bath gasifiers.

## Fixed bed gasification

### 1. The STEAG plant

The first, and so far the only, combined cycle power plant employing coal gasification is that installed by the West German utility Steinkohlen Elektrizitat AG (STEAG) at the Kellermann power station at Lünen (Bund et al., 1971). It uses fixed bed gasifiers and a supercharged combined cycle. The reasons for installing the plant, which has an electrical output of 170 MW were, according to STEAG (Krieb, 1974):

(1) development of power plant capable of meeting high environmental standards;

(2) improvement in efficiency;

(3) reduction of investment costs.

The total investment cost for this plant, construction of which started in 1969 and ended in 1972, was 80 million DM. The plant was said to be 15% cheaper than a conventional coal-fired plant with reheater.

The Lünen plant features five high-pressure Lurgi gasifiers. They can be charged with non-caking or slightly caking lump coal in sizes between 3 and 30 mm, and with a permissible undersize fraction of up to 7%. Ash contents up to 30% and water contents up to 15% are permissible, but the total content of incombustible matter must not exceed 35%. The gasifiers at the Lünen plant have an external diameter of 3·5 m and an overall height, including coal and ash locks, of approximately 20 m. Each gasifier has a coal throughput of 10-15 t h⁻¹.

The Lünen plant has a gross electrical output of 170 MW, made up of 96 MW from the steam turbine and 74 MW from the gas turbine. Although the total gas turbine output is 180 MW, 106 MW are required to drive the air compressor.

In order to match Lurgi gasifiers, which operate at 20 bar, to the supercharged boiler and gas turbine, which operate at 10 bar, it was necessary to install an expansion turbine and booster compressor. Details of the circuit are shown in simplified form in Fig. 8, together with the main technical data. The coal (a) is gasified in the gasifier (b) and the ash (c) is removed from the process. The gasifying media are steam (d) and air (e). The fuel gas (f) so produced leaves the gasifier at a temperature of 600°C, passes through the impingement scrubber and cooler (g) and enters the expansion turbine (h) at a pressure of approximately 20 bar and a temperature of

200°C. There, the pressure of the fuel gas is reduced to about 10 bar. The fuel gas then passes to the combustion chamber (i) in the pressurized boiler. The combustion gas (k) from the boiler enters the main gas turbine (l) at a temperature of 820°C. The waste gases are used for heating the feed-water in the feed-water preheater (o) and in the process are cooled to 170°C. The main gas turbine is coupled with and directly drives the air compressor (p). The compressed air is used as combustion air (q) for the combustion chamber and as gasifying air for the gasifier. The gasifying air is compressed in the booster compressor (r), which is driven by the expansion turbine, from approximately 10 bar, the pressure of the gas turbine cycle, to 20 bar, the pressure of the gasifying process.

Live steam (s) is generated in the pressurized boiler at 130 bar and 525°C to drive the conventional steam turbine set (t). The gasifying steam is bled from an extraction stage of the steam turbine (t).

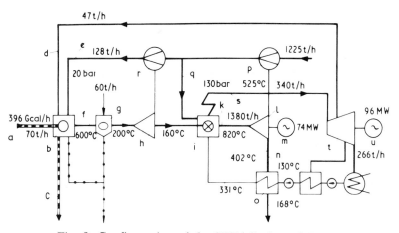

Fig. 8. Configuration of the STEAG plant, Lünen

This is an entirely new kind of plant and a number of problems could have been expected. As far as the gasifier is concerned, problems of control led to undesirable rises in the exit gas temperature, but it is now reported to be working satisfactorily. Although there have been several modifications to the gas cleaning equipment, the desired solids content for the gas has not yet been reached. However, operation of the combined cycle may be taken to be proved.

## 2. US Bureau of Mines

Mills (1972) has described how the US Bureau of Mines is investigating the pressure gasification of bituminous coals. In an attempt to overcome the limitations of conventional fixed-bed gasifiers with regard to size and caking capacity of the feed coal, a modified gasifier is being investigated at the Morgantown Energy Research Centre. The modification comprised a special stirrer which has vertical as well as rotary motion. Strongly caking coals have been successfully handled in a 1 m diameter unit at 18 t/day throughput.

## Fluidized bed system

### 1. Westinghouse

This project was started in 1972 when Westinghouse and their associates became convinced that there was a need for making low CV gas for power generation and also that fluidized gasification was the most promising route. The main item in the development programme is a process development unit, with a capacity of 0·5 t h⁻¹ coal (Archer et al., 1975).

The overall process scheme is shown in Fig. 9, in which multi-stage, fluidized-bed coal gasification is coupled with a combined cycle power plant. A two-stage, fluidized-bed process with sulphur capture is used to gasify coal with air and steam at temperatures up to 1150°C and pressures up to 20 bar. Particulate matter is removed from the hot gases which flow, without cooling, to the combustor where they are burned with excess air. The hot, burned gases expand through the turbine and then go to a waste heat system to produce steam which drives a steam turbine. About half the output of the gas turbine is used to drive the air compressor supplying the air for gasification and combustion; the other half is electrical power output. About an equal amount of electricity is also provided by the steam turbine.

In the process development unit, a detailed study is being made of the operation of the two stages labelled 'devolatilizer/desulphurizer' and 'combustor/gasifier' on the overall process scheme shown in Fig. 9. Westinghouse have stressed the point that their process is simply a novel arrangement of process steps, none of which is entirely new. Crushed, dried coal is fed into the devolatilizer/desulphurizer which is a recirculating, fluidized-bed reactor operating at 700 to 900°C. The fluidizing agent is the hot gas from the combustor-gasifier and is therefore a mixture of carbon monoxide, hydrogen and carbon dioxide diluted with nitrogen and undecomposed steam.

There are two features of note in this vessel. Firstly, it has a central draught tube and arrangements are made for gases to flow upwards through this tube at velocities of 5 m s⁻¹ or more. This imposes a deliberate circulation pattern on the fluidized bed which may be as high as 100 times the coal feed rate. The object of circulating this vast quantity of bed material is primarily to discourage agglomeration of the coal feed as it devolatilizes and passes for a brief interval through a phase when it is sticky. Because of the exceptionally high rates of heating, this sticky phase develops even with many coals normally regarded as non-caking. The aim is to surround each sticky coal particle with particles of devolatilized char or dolomite, both of which are

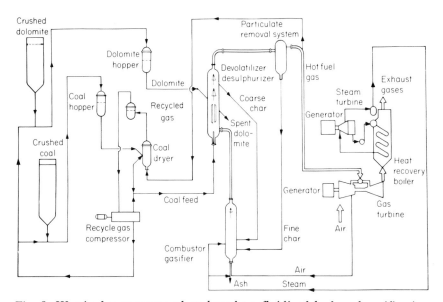

*Fig. 9. Westinghouse power plant based on fluidized bed coal gasification combined cycle.*

inert. A secondary advantage of the forced circulation is that the hot inlet gases are rapidly reduced from the gasifier temperature to that prevailing in the devolatilizer. The volatiles are driven off in an atmosphere containing hydrogen, which thus has the opportunity of reacting with the coal and char to form methane and release heat.

The second noteworthy feature of the reactor is the addition of dolomite. The aim is to fix the sulphur which, under reducing conditions, is evolved as hydrogen sulphide. On the flow sheet the spent dolomite and coarse char are shown as overflowing as two separate streams from the devolatilizer, but the means by which this is to be achieved are not given in detail.

Most of the heat required in the devolatilizer to raise the coal to 700-900°C is provided by the hot gases from the gasifier/combustor and by the solids carried over in them. Some heat is also obtained from hydrogasification; this can be supplemented if necessary by burning char in the downcomer around the draught tube.

The char from the devolatilizer/desulphurizer passes to the gasifier/combustor. This is another fluidized bed in which the gasifying agents are air and steam. The noteworthy feature of this step is that, although it is a fluidized bed, it is being operated with two distinct temperature zones. In the lower part of the vessel the char burns to produce carbon dioxide and steam and releases the major process heat requirements. Temperatures here range up to 1150°C, and many coal ash particles become sticky and agglomerate. The agglomerated ash then fails to fluidize, sinks to the bottom of the bed and can be removed. This neatly solves the problem of avoiding the loss of carbon normally ensuing from the need to maintain a high carbon concentration in the gasifier. The hot gases and solids circulating in the bed carry heat to the upper section where carbon dioxide and steam react endothermically with the char. In this way the required fuel gases are manufactured.

Westinghouse claim that this fluidized bed gasification process has the potential to overcome the limitations of other processes and provide an economic system for combined cycle power plants. Features of the process have been summarized by Archer *et al.* (1975) as follows:

(1) Utilizes wide variety of coals, including caking and high ash coals, and particle sizes, 6 mm × 0, without pretreatment; recirculating fluidized bed permits flexibility.

(2) Provides good heat economy; multi-stage operation permits selection of temperatures and residence time to achieve high efficiency; high temperature gas cleaning minimizes hot gas sensible heat losses.

(3) Prevents tar formation; maintaining high temperature fuel gas minimizes tar condensation.

(4) Provides high temperature sulphur removal; multi-stage permits temperature selection for utilization of limestone/dolomite for sulphur removal system.

(5) Provides necessary control for utility plant service; utilization of staged fluidized beds provides flexibility.

## 2. Institute of Gas Technology U-gas

The Institute of Gas Technology (IGT) is best known for its development of the Hygas process, which is a high pressure hydrogasification process using a series of fluidized beds. Recently, however, Loeding and Patel (1974) have given details of a process for making low CV gas called the U-gas process. Essentially this is a fluidized bed process operating at about 1050°C and 20 bar. It is proposed that the gas, after being cleaned by removal of particulates and sulphur compounds, may be used either in boilers, or as the fuel for a combined cycle.

The initial parts of the proposed process shown in Fig. 10 have much in common with other similar processes. Coal of any rank is crushed to below 6 mm, dried and fed through a lock-hopper system to a pretreater. This is a fluidized bed at system pressure in which the air, used as an oxidizing agent, destroys the caking properties of the feed coal. Surplus heat in the preheater is used to raise steam required in the process including that needed in the gasifier. The pretreatment step would be omitted for non-caking, sub-bituminous coals and lignite.

In the gasifier the coal is gasified using steam together with air. The off-gas from the pretreatment stage dilutes the gasifier off-gas. The temperature in the gasifier is about 1050°C, pressure about 20 bar and residence time about 45 minutes, while the fluidizing velocity is 0·3 to 0·8 m s⁻¹. As in the Westinghouse combustor/gasifier, the problem of selectively removing ash from a fluidized bed rich in carbon is solved by operating it to agglomerate the ash.

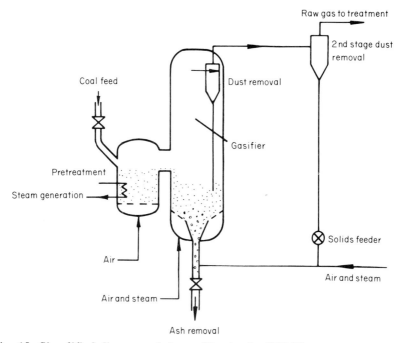

*Fig. 10. Simplified diagram of the gasifier in the IGT U-gas system.*

The base of the gasifier comprises an inverted cone or series of cones, and some of the steam and air used as a gasifying agent flows through a grid which slopes towards one of the cones. The remaining fluidizing gas enters the throat at the cone apex and jets upwards into the fluidized bed. By appropriate selection of operating conditions it is possible to arrange that in highly localized regions within the jet, instantaneous temperatures are achieved which are above the average bed temperature and high enough for the coal ash to soften. Ash particles then agglomerate and grow in size until they are too big to be fluidized even in the cone. They accumulate at the base of the gasifier and eventually find their way down the central off-take pipe.

It has been recognized that simply returning fines to the bed will lead to an unacceptably high solids loading in the cyclone system. Arrangements have therefore been made to burn the secondary fines and to use the hot gases thus created to heat and fluidize the material in the central off-take pipe.

Another feature worthy of note is the large disengaging space. The gas residence time here is 10-15 s which, at the prevailing temperatures, is sufficient to ensure that any tars formed are cracked. This leads to a great simplification of the processing steps for gas purification.

The test work done to date has been with a 1·3 m diameter unit operating at up to 2 bar pressure with air and steam. Encouraging results have been obtained: with a carbon content in the bed of 70%, the ash product had a size of 3-5 mm and a carbon content of 14%.

## 3. National Coal Board

The NCB is itself conducting a modest programme with financial support from the ECSC, on the manufacture of low CV gas. A fluidized bed system has been chosen because it is potentially less selective in coal types than fixed bed systems and avoids the high temperatures of entrainment systems which tend to volatile the alkalis in the ash.

The major problem with fluidized beds is the large carbon loss, particularly that suffered in the removal of ash. Processes like Westinghouse and U-gas aim to overcome this by agglomerating the ash and achieving a gravity separation. The NCB is interested in an alternative in which carbon-rich material removed from the gasifier is usefully employed in a separate fluidized bed combustor. A flow sheet of one such scheme is shown in Fig. 11.

## Entrainment systems

An example of the current work being done on entrainment systems is that by the Combustion Engineering Corporation in the US (Patterson, 1976); a simplified flow sheet is shown in Fig. 12. The gasifier has two distinct zones—a lower combustion zone and an upper gasification zone. Recycle char is burned in the combustor together with some additional coal to provide the heat required for gasification. Most of the ash is here melted and drawn off from the bottom of the gasifier as a molten slag. The hot combustion gases pass up into the reduction zone where they meet fresh coal and steam. By the time the gas reaches the top of the gasifier the temperature is down to 870°C and this is still further reduced by heat exchange equipment before the gas passes on to cleaning and purification processes.

A pilot plant of 5 t h$^{-1}$ capacity is currently under construction and this will be used to provide information required in the design of commercial scale gasifiers of this type.

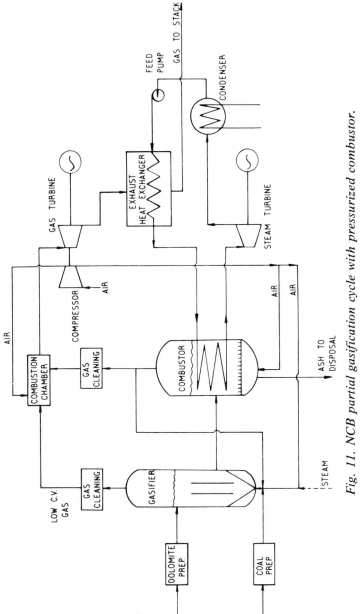

Fig. 11. NCB partial gasification cycle with pressurized combustor.

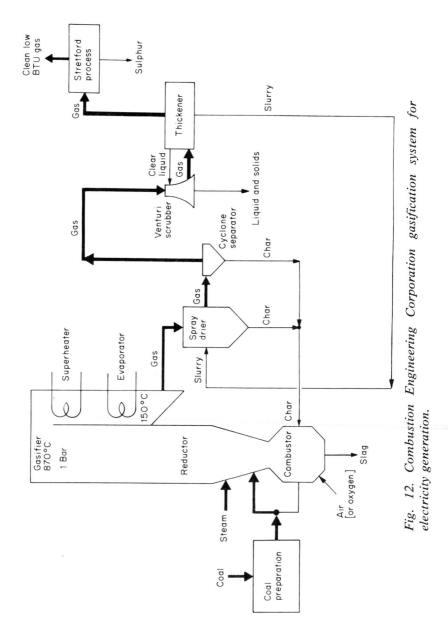

Fig. 12. Combustion Engineering Corporation gasification system for electricity generation.

## Molten bath gasification

Besides the Atgas or molten iron process described earlier, other workers in the US are investigating molten bath processes.

Atomics International (Trilling, 1974) are developing a process to produce sulphur-free gas to fire gas turbines and boilers, and a flow sheet of the proposed scheme is shown in Fig. 13. Coal and air are brought into contact in a bath of molten sodium carbonate which is useful in three ways:

(1) it acts as a solvent and heat reservoir for the reactants;

(2) the alkali catalyses gasification reactions, and high reaction rates are expected;

(3) the alkali fixes any sulphur.

The gasifier operates at 980°C at which temperature coal is almost completely gasified to give a product of which the fuel components are carbon monoxide and hydrogen.

The molten carbonate retains the coal ash and the sulphur so it must be continuously regenerated. It is first quenched and dissolved in water and then filtered to remove the ash. The filtrate is stripped of hydrogen sulphide using flue gas from the boiler and it is then carbonated, crystallized, centrifuged and dried before recycle to the gasifier.

It is claimed that the process can handle all kinds of coal (caking and non-caking), and that gasification rates are very high. The possible effect of any elutriated alkali on a gas turbine has not been discussed!

## VI. CHEMICAL SYNTHESIS

### The choice of gasification process

The foregoing section has discussed the production of low CV gas in which the presence of nitrogen is no great disadvantage, and in which air blown gasifiers are usually chosen. If, however, the gas is to be used for the production of hydrogen or chemicals, nitrogen is undesirable and the gasifier would be oxygen blown or be heated indirectly. Methane is another undesirable constituent which must be separated and used as a fuel or reformed at some stage. The necessity for methane processing adds to production costs and so a low methane concentration in the initial product is to be preferred.

Table III shows the product compositions of a number of gasifiers, some of which have already been described, and some of which have been developed primarily for SNG (pipe-line gas).

If a low proportion of methane is required, then the Lurgi, Molten-salt, Winkler and Koppers-Totzek processes are attractive. For carbon monoxide production where a low $H_2/CO$ ratio is desirable, then Atgas is the prime contender.

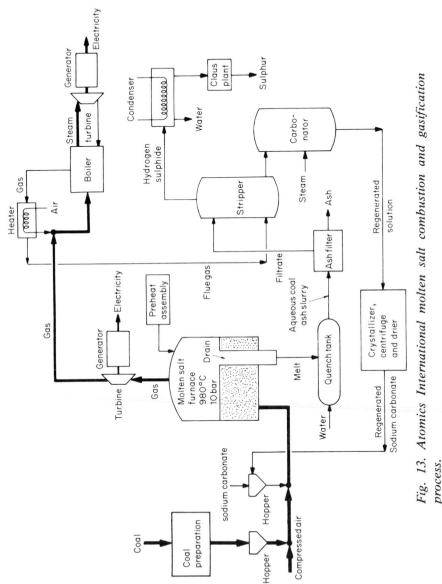

Fig. 13. Atomics International molten salt combustion and gasification process.

*Table III. Product compositions of various gasification processes.*

| Gasification Process | Raw gas composition (Mole %) | | | | |
|---|---|---|---|---|---|
| | $CH_4$ | CO | $H_2$ | $CO_2$ | $H_2/CO$ ratio |
| Hydrane | 73·2 | 3·9 | 22·9 | — | 5·9 |
| $CO_2$ — Acceptor | 20·9 | 17·0 | 53·8 | 6·6 | 3·2 |
| Lurgi | 9·4 | 18·5 | 40·4 | 29·5 | 2·2 |
| Synthane | 24·5 | 16·7 | 27·8 | 28·9 | 1·7 |
| Molten salt | 7·5 | 33·6 | 45·0 | 13·3 | 1·3 |
| Hygas | 18·7 | 23·8 | 30·2 | 24·5 | 1·3 |
| Winkler | 3·1 | 33·4 | 41·9 | 20·5 | 1·3 |
| Koppers-Totzek | — | 55·8 | 36·6 | 6·0 | 0·7 |
| Bi-gas | 15·6 | 44·0 | 24·4 | 14·0 | 0·6 |
| Atgas | 20·0 | 69·7 | 9·6 | — | 0·1 |

It is expected that hydrogen production from coal will assume considerable importance in the future. As supplies of petroleum become dearer and in shorter supply, processes for the liquefaction of coal, first for production of high-value aromatics and later for liquid fuels, will come into their own. These liquefaction processes, which are the subject of Chapter 8, will require large quantities of hydrogen, and by the time they reach the commercial stage it is unlikely that natural gas will be available for reforming. It will thus be necessary to make the hydrogen from coal feedstocks, and so gasification will be an important element in liquefaction processes. None of the gasification schemes listed is ideally suited to manufacture of hydrogen so recourse must be made to the shift and steam reforming reactions.

It should be noted that these comments about choice of processes have been based solely on the output compositions. Other factors such as efficiency, capital cost and reliability will naturally affect the choice.

## Pure products

Figure 14 outlines the broad steps required for the manufacture of chemicals. The raw product is purified by removing sulphur compounds and carbon dioxide and its composition is adjusted as required by reforming and shifting.

### 1. Hydrogen

Hydrogen may be obtained by the scheme shown in Fig. 15. The carbon monoxide is converted by the shift reaction with steam to hydrogen and carbon dioxide. The catalyst used for this reaction until recently has been an iron-chromium oxide mixture operating at about 30 bar and temperatures near 400°C. However, this catalyst is somewhat sensitive to sulphur compounds and is being replaced by cobalt-molybdenum catalysts. This enables

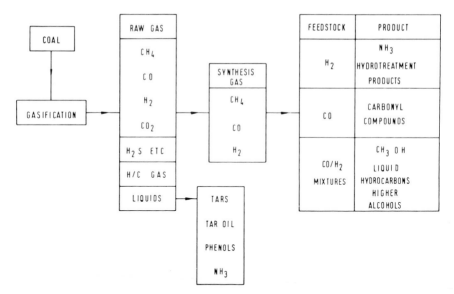

Fig. 14. Chemicals from coal gasification.

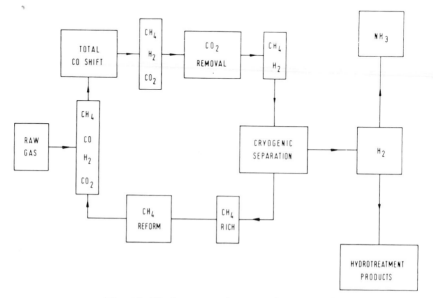

Fig. 15. Hydrogen and ammonia production.

the carbon monoxide shift to be carried out on the raw gas so that the sulphur compounds may be removed with the carbon dioxide in the following stage.

The boiling points of methane (109 K) and hydrogen (21 K) allow relatively easy cryogenic separation. The methane-rich stream may then be used as a fuel or recycled after passing through a reforming stage.

In methane reforming, steam is used at 750°C at pressures up to 20 bar, with a nickel-based catalyst. $H_2/CO$ ratios up to 5:1 are obtained.

The hydrogen thus produced is sufficiently pure for many industrial purposes, but if it is to be used in ammonia production it is necessary to remove all oxygen-containing compounds such as carbon monoxide and dioxide (by adsorption or washing with liquid nitrogen).

## 2. Carbon monoxide

The second main component of synthesis gas, carbon monoxide, may be separated as a pure gas using either cryogenic separation or a newly developed adsorption technique, the COSORB process reported by Haase and Walker (1974). The latter uses a solution of cuprous aluminium chloride in an aromatic base to remove carbon monoxide from the gas stream by formation of a complex. The tail gas consists of a mixture of methane and hydrogen which may either be separated or used as a fuel gas. The carbon monoxide is recovered as a pure ($\sim$99%) gas which may then be used as a chemical feedstock.

In a pure form, carbon monoxide is used to produce acids, esters, acrylic acid, hydroxyl acids, formamides and metal carbonyls. Some of these products are themselves converted to higher esters, ketones, aldehydes, glycols and plasticizers. Many of the metal carbonyls, of which nickel is an example, are converted into powdered metals of high purity. Some examples of reactions are given in Table IV.

*Table IV. Reactions of carbon monoxide.*

| Reactant | Example | Product |
|---|---|---|
| Alcohol | $CH_3OH + CO \leftrightharpoons HCOOCH_3$<br>(alkali metal alkoxide catalyst)<br>$CH_3OH + CO \leftrightharpoons CH_3COOH$<br>(acidic catalyst) | Ester<br><br>Acid |
| Aldehyde | $HCHO + H_2O + CO \rightarrow CH_2OH.COOH$ | Hydroxy acid |
| Acetal | $CH_3OCH_2OCH_3 + CO \rightarrow CH_2(OCH_3)COOCH_3$ | Methoxy ester |
| Amine | $(CH_3)_2NH + CO \rightarrow HCON(CH_3)_2$ | Formamide |
| Acetylene | $C_2H_2 + H_2O + CO \rightarrow CH_2CHCOOH$ | Acrylic acid |
| Chlorine | $Cl_2 + CO \rightarrow COCl_2$ | Phosgene |
| Metal | $Ni + 4CO \leftrightharpoons Ni(CO)_4$ | Metal carbonyl |

## 3. Methanol

The reaction sequence, which is similar in some respects to that for hydrogen, is shown in Fig. 16. Carbon dioxide is used in methanol synthesis to help control the reaction temperature since its heat of reaction with hydrogen is only about half that of carbon monoxide and in addition, the water produced helps quench the reaction. Carbon dioxide is thus removed in the stage following the shift reaction in a controlled way. The composition of the feed gas to the methanol synthesis loop is such that the ratio $H_2$: (CO + 1·5 $CO_2$) is 2:1. The synthesis must be carried out under pressure; the most favoured process operates at pressures of 50 to 100 bar. The catalyst is copper-based and the temperature used is about 250 to 300°C. The crude methanol is condensed and distilled while unreacted gases are recycled in the synthesis loop. A purge stream of methane-containing gas is either used as fuel or reformed.

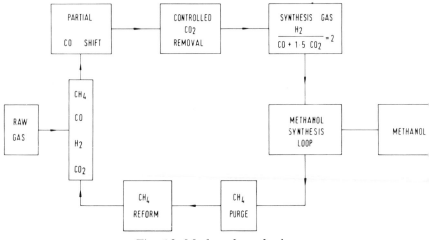

*Fig. 16. Methanol synthesis.*

## 4. Other synthesis products (Fischer-Tropsch)

With the two exceptions of methane and methanol, which can be formed without significant quantities of by-products, the reaction of carbon monoxide with hydrogen yields a complex mixture of hydrocarbons, alcohols, carbonyl compounds and acids. The distribution of products depends to a considerable extent on the reaction conditions, i.e. $H_2$:CO ratio, temperature, pressure, catalyst, reactor systems, etc. The role of Fischer-Tropsch synthesis in coal liquefaction is described in Chapter 8.

## VII. SUBSTITUTE NATURAL GAS

Natural gas, which consists mostly of methane and has a calorific value of 37 MJ m⁻³, is an important component of the energy economy of most of the industrialized countries of the Northern Hemisphere. It is an attractive fuel for the premium markets of space-heating and specialized industrial applications, and the development of high-pressure, underground pipeline systems has permitted the transport of energy at very high density in a manner which makes little adverse impact on the environment. In the United Kingdom, for example, these attractive features have led to a rapid exploitation of the natural gas reserves in the North Sea. The reserves are finite, and even if the use of gas is confined as at present to premium markets, it is likely that supplies will begin to decline at some time near the turn of the century.

The large investment in the distribution system, and the advantages of transmitting large quantities of energy in this form, argue in favour of the production of a substitute gas of high calorific value which can sustain the supply as the natural sources gradually dwindle. Probably the simplest technical solution would be to manufacture methane from oil-based feedstocks, but it is likely that shortages in the supply of oil will occur at the same time as, or even before, those of natural gas. Reserves of coal exceed those of gas and oil, both globally and locally in Western Europe, and it is therefore coal which is the preferred source of substitute natural gas (SNG). According to recent forecasts (Dept of Energy, 1978), SNG production may become the most important use of UK coal in the first two decades of the next century.

The first stage in the production of SNG from coal is the manufacture of a synthesis gas in a gasifier which is oxygen blown or which uses indirect heating. It is arguable that it is desirable to maximize the production of methane in this first stage (see Table III), and some systems now under development have this objective. It is likely, however, that the choice of gasifier in a commercial system will depend more on reliability in operation than on the proportion of methane in the off-gas. In any case, the methane content of the primary products rarely exceeds 15%. The problem, then, is to convert a mixture of hydrogen and carbon oxides to methane.

The appropriate reactions between hydrogen and carbon oxides can be promoted by catalysts, usually nickel-based, but it can be difficult to achieve completion. It would be necessary to remove residual oxides of carbon from the product gas, and this would be relatively easy with carbon dioxide, but less so with carbon monoxide. Moreover, the presence of sulphur compounds in the feed gas would tend to poison the catalyst.

The preferred method is first to subject the raw gas from the gasifier to the shift reaction in order to reduce the amount of carbon monoxide and produce more hydrogen. It is next scrubbed, carbon dioxide and sulphur compounds being removed.

The gas then enters the stage where methanation is promoted by the nickel-based catalyst. The difficulty here is one of temperature control, since methanation reactions are strongly exothermic. The optimum temperature is about 350°C, and if the temperature rises too high, the catalyst is deactivated. On the other hand, if the temperature falls too low, carbon monoxide may react with the catalyst to form nickel carbonyl. Temperature control is achieved by performing methanation in a number of stages, and also by recycle of cold product gas.

SNG production from coal has been demonstrated on the commercial scale by British Gas at their pilot plant at Westfield, Scotland, using synthesis gas from a Lurgi gasifier. Although this was merely a large experiment, the scale and success of the operation may be judged from the fact that the product was routed to the distribution network in place of natural gas, and that no consumer noticed any difference (Gray, 1976).

## VIII. CONCLUSION

The gasification processes and applications described in this chapter represent a few examples from a wide range of development work in progress. The variety of concepts under study is evidence of the renaissance of interest in coal gasification and its rapidly increasing importance in energy technology.

## REFERENCES

Archer, D. W., Keairns, D. L. and Vidt, E. J. (1975). *Energy Comm.* **1**, 115—134.
Bund, K., Henney, K. A. and Krieb, K. H. (1971). World Energy Conference, Bucharest. Paper 2, pp. 3-71.
Department of Energy (1978). 'Energy Forecasts.' Energy Commission, Paper 5.
Foster, L. D. (1970). "Gas Turbine Fuels." ESDA-7004, General Electric.
Fraley, L. D. and Kumar, C. A. (1975). "Clean Fuels from Coal." Symposium II, IGT, Chicago.
Giramonti, A. J. and Lessard, R. D. (1975). *Appl. Energy,* **1**, 311.
Gray, J. A. (1976). *Coal Energy Q.* **11**, 15-22.
Haase, D. J. and Walker, D. G. (1974). *Chem. Eng. Prog.* **70**, 74.
Krieb, K. H. (1974). *Chem. Econ. Eng. Rev.* **6**, 23-27, 58.
La Rosa, P. and McGarvey, R. J. (1975). "Clean Fuels from Coal." Symposium II, IGT, Chicago.
Loeding, J. W. and Patel, J. G. (1974). 67th Annual Meeting of the American Institute of Chemical Engineers. Paper 58b.
Mills, G. A. (1972). Proc. 4th SPG Symposium, Chicago.
Patterson, R. C. (1976). *Combustion* **47**, 28-34.
Trilling, C. A. (1974). ASME Paper 74-WA/PWR-11.

# 8 Liquefaction of Coal

## J. OWEN

*Deputy Director, Coal Research Establishment, National Coal Board*

## I. INTRODUCTION

The long-term incentive to make liquids from coal is very great. World petroleum reserves can only be reckoned in decades, at present rates of consumption, while coal reserves are likely to be available for centuries. Therefore it is important to develop those coal conversion processes which will provide the essential liquids that at present can only be obtained from petroleum. In this context, the essential liquids are those required for transport fuels and chemical feedstocks. This being so, the production of a coal-based refinery feedstock, i.e. a crude oil replacement, represents a large future industry. This does not imply that coal-based feedstocks will suddenly replace crude oil, but that, as crude oil supplies dwindle, coal will gradually take over. The detailed patterns will differ in different countries. In the US there is a strong incentive to replace fuel oil with a clean, coal-based alternative oil. That is unlikely to happen in Europe as it is too expensive to convert coal for this use; it is more sensible for us to burn coal directly. Thus, in Europe, and especially in the UK, liquefied coal is likely to be made for transport fuels and chemical feedstocks.

Previous chapters have included discussion on the physical and chemical structures of coal, in which it has been stated that most coals soften and melt at temperatures of about 400°C. Some petrographic constituents become more fluid than others, while the inertinite group of macerals does not soften at all. Many attempts have been made to separate out the more liquid fractions of the decomposing coal as they are of higher potential value, being higher in hydrogen content. Figure 1 shows various fuels arranged in order of

163

ascending hydrogen to carbon ratio. All coals, and their components
(macerals) are near the bottom; petroleum oils are in the middle, while the
hydrocarbon gases are at the top.

Solvents are used to liquefy coal but, without heating, the yield of
extractable material from coals is very low, even when using the strongest
solvents. Heat treatment at 350°C and above produces components which
are extractable and causes thermal degradation with reduction in average
molecular weights, which also improves extraction. (There is a short account
of the actions of different solvents in Chapter 1.) It is the objective of many
processes to stabilize and separate these species before they repolymerize and
thus the medium in which the thermal treatment occurs can affect the
products considerably.

The review that follows will describe briefly several different approaches
adopted throughout the world.

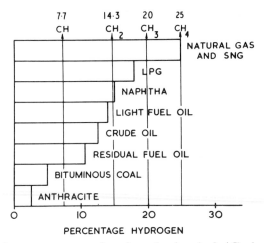

*Fig. 1. Hydrogen contents of coals and other fuels (Grainger, 1978).*

## II. HEAT TREATMENT IN A GASEOUS MEDIUM

### Gasification and synthesis

In Chapter 7 it has been shown that coal can be gasified to synthesis gas, i.e.
carbon monoxide and hydrogen, which can be converted to liquids by using
suitable catalysts and processing conditions in the Fischer-Tropsch synthesis.
This is being operated by Sasol in South Africa, where it is politically
important to have a measure of independence from imported oil. Further-

more, when the Sasol plant was first built, coal there was extremely cheap as it lay mainly in thick seams close to the surface. Coal is gasified in a conventional Lurgi gasifier (i.e. fixed bed, rotating grate) and the crude gases are purified and subjected to the water gas shift reaction which produces the correct ratio of hydrogen to carbon monoxide. This part of the process is not shown in Fig. 2 which begins at the gas reforming stage. The Sasol/Kellogg process differs from its predecessors in being a fluidized-bed catalyst synthesis. The remaining (standard) stages are concerned with separation and purification.

This process reduces the coal to very small units (carbon monoxide and hydrogen) and then recombines them into hydrocarbons and alcohols. It is not very efficient on a thermal basis nor is it very cheap. Thus, it is not in use anywhere else in the world today.

## Pyrolysis

In pyrolysis, coal is heated in the absence of air and much of the volatile matter, released from coal during heating as in coking plants, is collected as liquids on quenching. If coal is heated slowly then the volatiles are evolved slowly, extensive secondary decomposition takes place and there are only small yields of liquid produced. The more rapidly the coal is heated, the larger the proportion that volatilizes. At very high heating rates ($10^5$ °C s$^{-1}$), up to 70% of a coal can be volatilized, much of it as pitch, even though the standard laboratory volatile matter test shows only 40% for the same coal (Kimber and Gray, 1967).

In the COGAS process, (Chap. 7, Fig. 4) coal is injected into a multi-stage fluidized bed, which gives a fairly high heating rate, and there is an increased production of liquids. There are several subsequent stages in which the solid char is gasified, the primary tar is treated with hydrogen and the gases are reformed to substitute natural gas. This process has been under investigation for several years and a demonstration was carried out some time ago, using hydrogenated coal liquids to fuel a US destroyer.

Another pyrolysis process is the Garrett system which claims to maximize the liquid yields by using a disperse phase (entrainment) reactor to give high coal heating rates with recycle of reheated char.

## Supercritical extraction

This subject has also been referred to in Chapter 1. Recapitulating, fluid extracts are obtained by treating coal with selected solvents at temperatures above their critical points. Under supercritical conditions the solvent extracts some of the coal substance and carries it from the reactor. Reduction of the pressure then causes almost all of the extract to precipitate and there is little gas formed. The NCB is pioneering this technology (Fig. 3) and liquid yields

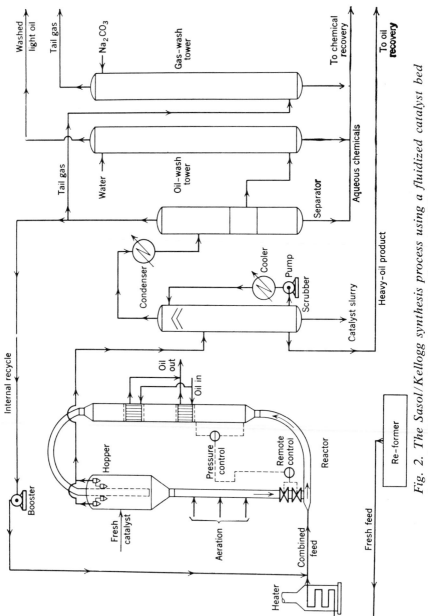

*Fig. 2. The Sasol/Kellogg synthesis process using a fluidized catalyst bed (reproduced by permission from Kirk-Othmer, 1964).*

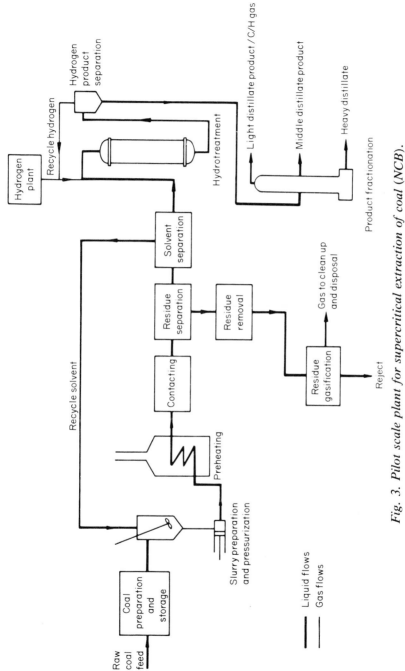

Fig. 3. Pilot scale plant for supercritical extraction of coal (NCB).

of 30% wt have been obtained; these can be raised to 50% by simultaneous catalytic hydrogenation and extraction.

All the above processes rely on thermal treatment to break bonds in the coal structure, followed by outward diffusion and/or active extraction of the volatile products thereafter.

## III. HEAT TREATMENT IN A LIQUID MEDIUM

### Liquid solvent extraction without added hydrogen gas

In the simplest case, the solvent acts in a physical manner, dissolving out much of the coal substance and the solvent can later be recovered, virtually unchanged, by distillation. The best solvents produce extracts containing about 90% of the original coal substance, though this is not entirely a physical process. Using coal tar oils, which are among the best solvents, at temperatures above 400°C, most of the coal substance is dissolved with the formation of only a small percentage of gas. The average molecular weight of the dissolved coal is high (average over 1000), there being little material soluble in benzene although usually it is all soluble in quinoline. The undissolved coal is mainly mineral matter and fusain. It should be understood that coal is dissolved in the solvent and is not present as a solid suspension as, for example, in a slurry. Different liquid solvents dissolve coal to varying degrees, the most efficient being naphthenic (i.e. saturated aromatic) compounds. The ability to transfer hydrogen from the solvent to the coal substance, i.e. to act as a hydrogen donor, is especially important as transferable hydrogen prevents repolymerization (Section IVff.). Extracts produced in this way are liquids or low melting point solids, which are especially amenable to subsequent chemical treatments. In many ways the extracts may be regarded as analogous to crude oil, though the chemical compositions are different. At this stage, apart from the relatively high costs (in Europe), the extracts are not sufficiently "clean" for use as fuel oils.

One possible commercial application of this solution process is for the production of high purity carbons, in which extracts are coked and graphitized in various ways to make electrodes for the steel and aluminium industries and carbon fibres, as indicated in Chapter 9. In the intermediate coking stage associated with carbon production, sufficient solvent oil is regenerated to make the process self-supporting in solvent, this being an essential requirement for any large new industry. The NCB is especially interested in this system for making high purity carbons and graphites and experiments are currently being carried out with a 0·5 t/day plant to provide design data for a demonstration unit to make 20 t/day of electrode coke suitable for the arc steel industry (see Fig. 4).

For further specialized information on electrode coke, the reader is

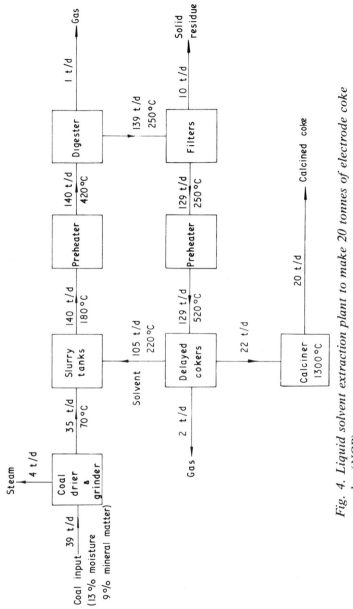

Fig. 4. Liquid solvent extraction plant to make 20 tonnes of electrode coke per day (NCB).

referred to papers presented at a recent conference on 'Tar, Pitch, Solvent Refined Coal and Petroleum, as Used in Carbon and Graphite Production.' (Marsh et al., 1978).

Solvents of higher hydrogen donor power are not required for carbon production, but their use is beneficial in making extracts especially for subsequent downstream refining (e.g. hydrocracking) as will be described subsequently.

## Solvent extraction with added hydrogen

### 1. Without catalysts

As indicated in Fig. 1, a far reaching and fundamental difference between coal and crude petroleum is in the carbon/hydrogen ratios. Thus, when the aim is to produce usable liquids from coal then hydrogen must be added and there are various ways of proceeding.

Though many solvent extraction processes have been proposed and tried, none has survived commercially. However, since the oil crisis of 1973, substantial funds have been made available, largely by Governments and the oil companies, to develop processes for conversion of coal to oils. First of all, those processes where hydrogen is added in one form or another during the solvent extraction process will be described.

Historically, one of the best known industrial plants was the Pott-Broche process used in Germany during World War II. Extraction conditions were 415-430°C for 60 minutes at 100-150 bar with an anthracene oil solvent derived from coal tar, containing a proportion of hydrogen donor compounds. Fresh solvent was added to a recycle stream in order to maintain the required solvent power, although an alternative sometimes employed was to add hydrogen gas during digestion. The plant had a capacity of 30 000-50 000 t/year, but was closed shortly after the war because of practical difficulties and expense associated with the ceramic filter candles employed.

In the early 1960s, the Spencer Oil Company (US) started investigations aimed at the production of Solvent Refined Coal by digestion with solvents under hydrogen gas pressure. The object was to reduce the inorganic and organic sulphur contents of the high sulphur US coals to environmentally acceptable levels for combustion in power generating plant. Later, a plant was built for a consortium headed by Southern Services Utility Company at Wilsonville in Alabama, and has been successfully processing 6 t/day of coal since 1974.

At Fort Lewis, near Seattle, Gulf Oil, with the financial support of the US Government Department of Energy, built and now operates a 50 t/day plant using a similar process. No catalyst is added and conversions of over 90% of the coal substance to liquids are obtained. The process flow sheet (Fig. 5) is similar to that for the Wilsonville unit and different types of process equipment are being tested. Recent emphasis has been on elimination of filters, these being replaced by a vacuum distillation unit.

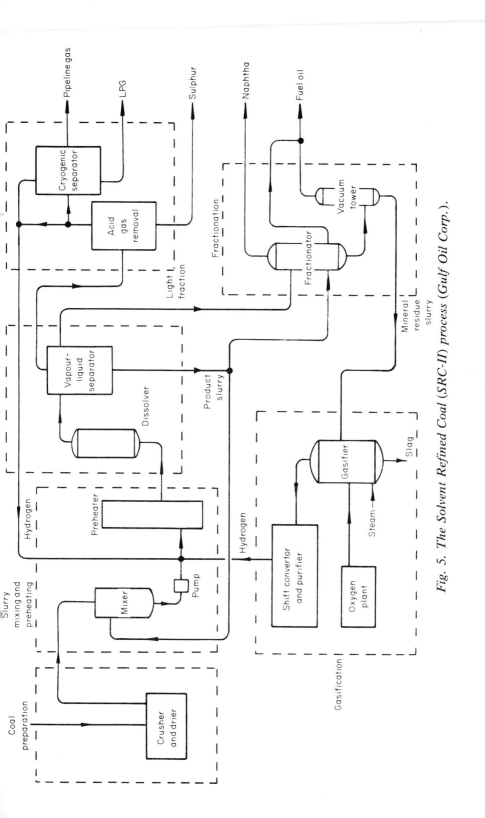

Fig. 5. The Solvent Refined Coal (SRC-II) process (Gulf Oil Corp.).

Gulf have shown, by testing coals with varying amounts of mineral matter, that some minerals in coal have beneficial catalytic action on the hydrogenation reactions. The process has a net hydrogen consumption of about 2% weight on coal and gives extracts with melting points of 100-200°C, which is lower than those of non-hydrogenated extracts. Moreover, extracts with sulphur contents of less than 0·5% can be made even from high sulphur coals.

Similar, but smaller, research units exist in other parts of the world, for example, Sasol in South Africa, Bergbau-Forschung in Germany and others in Poland and Japan.

Another liquefaction process with no added catalyst (Fig. 6) is the COSTEAM one in which a 50% coal and water slurry reacts in synthesis gas (CO + H₂) instead of pure hydrogen, in a stirred reactor at 300 bar and 420°C. The residence time is 1 h and water is added to bring the steam/carbon ratio to 0·66 : 1. This process has been developed at the Pittsburgh Energy Technology Centre specifically for use with lignites and a 50 kg/day coal plant is now being tested.

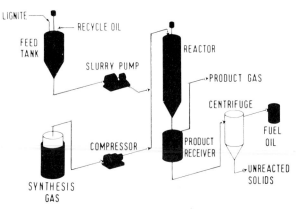

*Fig. 6. The COSTEAM process for conversion of lignite to low sulphur fuel oil (Pittsburgh Energy Technology Center, U.S. Dept. of Energy).*

The Clean-coke process which is being developed by the United States Steel Corporation also falls into this class. It is based on work done in the early 1960s at the corporation's Monroeville laboratories, aimed at finding a suitable process for exploiting high sulphur coal and making gaseous, liquid and solid products for a wide variety of outlets. After initial preparation, the coal is divided into two roughly equal portions of which one is carbonized to produce low-sulphur char, tar and hydrogen-rich gas, and the other hydrogenated to a chemical-rich liquid and gas. The products are separated from the residue of unconverted coal and ash by flash evaporation. Liquids from all operations are combined and processed to provide chemical and fuel feedstocks. Similarly, all gas fractions are combined and processed to provide various industrial gases, ammonia and fuel gas, and the flow sheet shows a sulphur plant. Char is blended with binder and calcined to produce low-sulphur formed coke for metallurgical use.

## 2. With catalysts

Where more than 2% hydrogen is to be added, i.e. when distillable hydrocarbons are required, catalytic hydrotreatment is required. Two of the best known processes are H-coal and Synthoil.*

In the former, developed by Hydrocarbon Research Inc., USA, a slurry of coal and solvent oil is passed into an 'ebullating' bed of catalyst, meaning a violently fluidizing system with a central recycle draft tube which promotes downward circulation of the burden as shown in Fig. 7. Much experimental

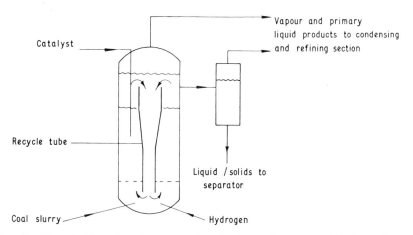

*Fig. 7. The ebullated bed reactor in the H-coal process (Hydrocarbon Research Inc.).*

work has already been completed using a 3 t/day plant and enough has been learned to design a commercial-sized unit. The degree of hydrogenation can be varied and the H-coal reactor is run either to give a low sulphur fuel oil, or to make a syncrude more suited to further refining. A 600 t/day demonstration plant is now being built in Kentucky.

In the Synthoil* process, developed by the US Bureau of Mines, now part of the US Department of Energy, the catalyst bed is fixed and a coal-solvent slurry is pumped into the reactor with hydrogen at 450°C and pressure at

*This name is no longer used.

100-300 bar, as shown in Fig. 8. A high degree of turbulence is induced to minimize deposition of carbon and coal mineral matter. A 10 t/day pilot plant is currently under construction.

Similar yields are obtained in both the Synthoil and the H-coal processes, i.e. 3 barrels of low sulphur oil per tonne of coal (equivalent to 50% wt, 65% thermal). Thermal efficiencies are high and costs are said to be low.

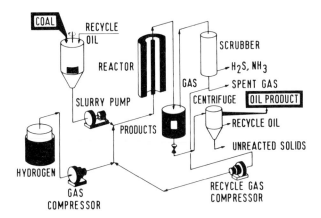

*Fig. 8. The SYNTHOIL process for conversion of coal to low sulphur fuel oil (Pittsburgh Energy Technology Center, U.S. Dept. of Energy).*

## Processes with separate hydrogen treatment

In these systems, the solvent is hydrogenated in a separate vessel to increase hydrogen donor power and thus it carries extra hydrogen back into the extraction stage. Simultaneous hydrogenation from solvent and hydrogen gas when present occurs during extraction but there is always significant hydrogenation from the solvent. The relative contributions of hydrogen from the gas and via the solvent depend on the operating system. The recycling solvent builds up the hydrogen donor level and this usually gives better, i.e. more saturated, liquid products.

The Exxon process, shown diagrammatically in Fig. 9, utilizes a separate solvent hydrogenation step and high pressure hydrogen gas is present in the digestor as well. The Exxon system also dispenses with filtration and substitutes a vacuum distillation tower, the residue from which is taken to a delayed coker to maximize liquid yields. Coke from the coker has over 50% ash and is gasified to supply hydrogen. The process is completely self-contained and produces about 40-45% of distillates. A demonstration plant of 250 t/day coal input, jointly financed by the US Department of Energy and a consortium led by Exxon, is now under construction.

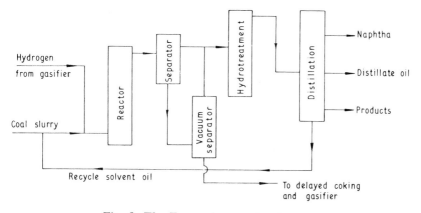

*Fig. 9. The Exxon donor solvent process.*

In the NCB Liquid Solvent and Hydrocracking Process, a hydrogen donor solvent (hydrogenated recycle oil) is used in a process similar to that shown in Fig. 4, to maximize the production of primary extract. The extract is filtered and then hydrocracked over special catalysts to give maximum yields of oils boiling below 400°C. Higher boiling oils, which have also been hydrogenated during the hydrocracking reaction are recycled to the solvent extraction stage as shown in Fig. 10. In this process, up to 90% of the coal substance is converted and product distillates have been shown to be useful as chemical feedstocks for aromatics production and as good blending stocks for internal combustion engines.

Both the NCB liquid solvent and supercritical gas extraction processes have been accepted by the UK government for development to a larger scale and plans for 1 t h⁻¹ pilot plants are being prepared.

Most of the solvent extraction processes mentioned so far make fuel oil as a main product, although the use of catalysts permits a higher degree of hydrogenation than is necessary for fuel oil alone. It is usually better to hydrocrack coal extracts after the mineral matter is removed as many catalysts are easily poisoned.

Several organizations are testing and developing catalysts for this work and there is much relevant experience available from the hydrocracking of petroleum fractions. It is a question of developing those best suited to coal-based materials.

## Direct hydrogenation

One of the first coal hydrogenation processes was that pioneered by Bergius in Germany in 1913, and extensively developed both in Germany and by ICI at Billingham in the 1930s. In this process a paste of coal and oil was directly

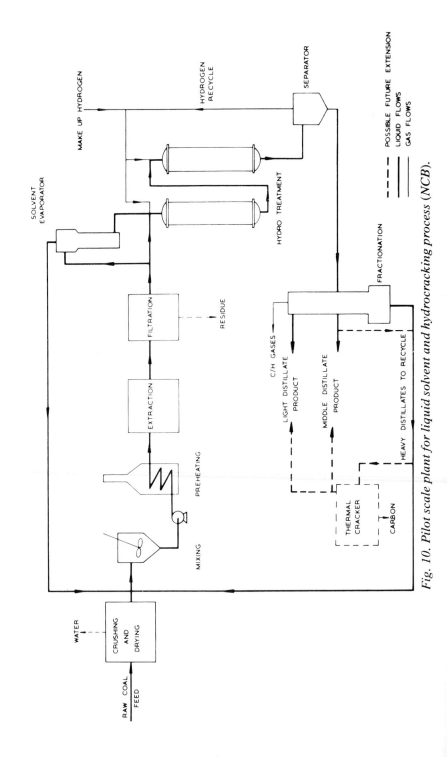

Fig. 10. Pilot scale plant for liquid solvent and hydrocracking process (NCB).

hydrogenated with very high pressure hydrogen, promoted by a cheap, once through, expendable iron oxide catalyst. This was done on a very large scale during World War II in Germany when millions of tonnes of hydrocarbon oils were made from coal as Germany was cut off from the most of the outside supplies of petroleum. However, the low grade catalysts necessitated using very high temperatures and high hydrogen pressures (700 bar), so the process was expensive and mechanically troublesome and it was abandoned immediately after the war.

The ICI developments in the 1930s concentrated on using better catalysts, e.g. tungsten sulphide, in fixed bed reactors for hydrogenating tar oils; direct conversion of coal was soon abandoned as being too difficult and expensive.

## IV. REACTION MECHANISMS OF COAL LIQUEFACTION

Having reviewed some liquefaction processes, it is useful to consider the physical and chemical changes that occur. In a limited space, this cannot be done extensively and it is best to concentrate on one or two aspects in some detail.

### Extraction efficiencies

The use of liquid solvents is common to most of the processes described, especially those that give the highest conversion to liquids. The most common solvents are derived from coal, usually fractions of tar produced as byproducts from the carbonization of coal for metallurgical coke. In many cases the original solvent is modified somewhat during processing, e.g. by hydrogenation and by thermal degradation of the coal, into useful process solvent. It is instructive to see why anthracene oil makes such a good coal solvent and some investigational work has been carried out at the NCB Coal Research Establishment.

Anthracene oil, a coal tar cut boiling between 200 and 450°C (containing over 100 known compounds—see Fig. 11) was distilled and fractionated narrowly so there were relatively few pure components in each fraction (see, for example, Fig 12). Each cut was then tested for solution power, in a small bomb, to determine the proportions of dissolved and undissolved coal after standard heat treatment.

Figure 13 shows the results for solution power of distillation cuts from both fresh and hydrogenated oil fractions. For the anthracene oil, the peaks show clear dependence of solvent efficiency on composition rather than on boiling point. Looking at the hydrogenated cuts, however, shows that the efficiencies of extraction, especially of the poorer components, increased markedly throughout the entire boiling range. The total extraction is improved because the number of components which act as hydrogen donors has been increased, a finding of far reaching practical importance.

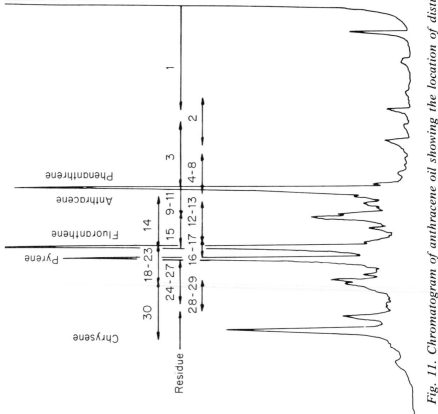

Fig. 11. Chromatogram of anthracene oil showing the location of distillate fractions.

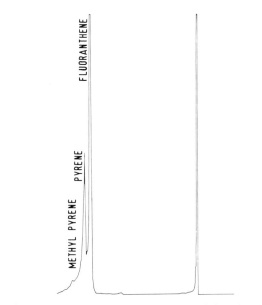

Fig. 12. Chromatogram of anthracene oil fraction No. 20.

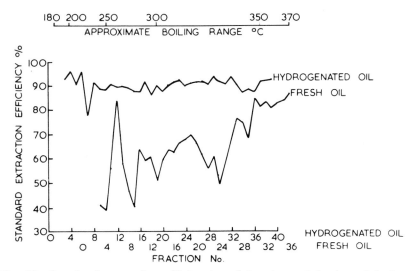

Fig. 13. Standard extraction efficiencies of fractions of fresh and hydrogenated oil.

## Viscosity changes

The reason for interest in viscosity is a practical one because of the important effect it has on one of the major processing unit operations, i.e. solids separation, the rate of filtration being inversely proportional to the filtrate viscosity. It should be understood that the viscosities referred to are always those of the reacting mixtures of coal and solvent at whichever stage of the process is being considered.

In practice, viscosity is not the only important factor and the resistivity of the filter cake has to be optimized against the operating requirements for a minimum viscosity. Much progress in making filtration a practical procedure has been accomplished by choosing conditions which balance cake resistivity, dependent on particle size distribution and physical consistency, against solution viscosity.

There are interesting viscosity changes occurring with temperature and time for coal extracts made with anthracene oil. The general form for changes in viscosity with time at constant temperature is shown in Fig. 14, where there are four distinct zones separated by three points of inflexion. The following explanation is a working hypothesis.

First, there is a steep rise in viscosity which takes place within seconds at all temperatures studied, reaching a peak within a few minutes. Coal swells prior to disintegration and this contributes to the measured viscosity increase.

After reaching the first peak, the viscosity drops for 20-30 minutes, indicating a zone of depolymerization. The minimum viscosity reached between C and D and the maximum between D and E both occur at times which are virtually independent of temperature, i.e. after 30 minutes and 1 hour respectively. Simultaneously, polymerization reactions occur, causing viscosity increases before the end of depolymerization, and it is the combination of the two reactions and their relative rates that controls the position of minimum viscosity.

Viscosities are not greatly altered between 380 and 420°C, indicating a low activation energy which is generally consistent with polymerization reactions. The zone at E shows steadily reducing viscosity which could be a consequence of the coking of the coal in solution. When this reaction has proceeded to completion, the digest would have almost the same solids content as the original coal-solvent mixture at the beginning and thus the same solvent viscosity would be expected. This is observed in practice. However, in the early stages of zone E, the reaction occurring is probably not the production of a real coke, but rather a locking of large polymers by cross-linking or the ordering into liquid crystal structures. Either of these mechanisms would result in a viscosity decrease.

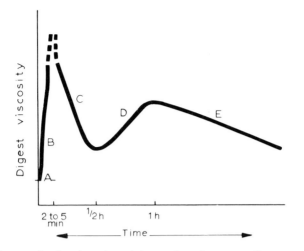

*Fig. 14. Changes in the viscosity of the coal-anthracene oil system with time.*

## V. CONCLUSION

The underlying problem in liquefying coal is to increase the hydrogen to carbon ratio and to do this cheaply. There have been many attempts, none commercially successful as yet. Until the oil crisis of 1973, no process was known which could convert coal into oil at anything like a competitive price, but the improvements in petroleum technology of the last 20 years, and the huge jump in oil prices, have brought economic coal conversion processes much nearer. Moreover, with the oil reserves being so much smaller than those of coal, it is essential to develop economic conversion processes before crude oil disappears. The very long lead times needed to bring new processes into large scale production mean that the requisite technologies should be developed now.

## REFERENCES

Grainger, L. (1978). *Chem. and Ind.* 12-15.
Kimber, G. M. and Gray, M. D. (1967). *Combust. Flame* **11,** 360-362.
Kirk-Othmer (1964). Encylopedia of Chemical Technology (2nd ed.) Vol. 4, pp. 446-489. Wiley Interscience, New York.
Marsh, H. *et al.* (1978). *Fuel, Lond.* **57,** 130-174.

# 9 The Coal Tar Industry and New Products from Coal

## J. OWEN

*Deputy Director, Coal Research Establishment, National Coal Board*

## I. INTRODUCTION

In Chapter 3, the coal carbonization industry was dealt with in terms of coke, the primary product. However, another large industry exists to deal with the secondary or by-products produced during coal carbonization. Broadly, these products are coke oven gas, ammonia liquor, crude benzole and tar, and typical yields from high and low temperature carbonization are shown in Table I. In 1976, about 875 000 tonnes of coal tar and 230m litres of benzole were produced in the UK.

It is often claim that the organic chemical industry had its origins in using the products of benzole and coal tar refining during the nineteenth century, notably in making the first synthetic dyestuffs. Tar processing grew and

*Table I. Yields of bulk products from high and low temperature carbonization of coal (%weight on dry coal).*

| Product | Low temperature | High temperature |
|---------|-----------------|------------------|
| Gas | 7·6 | 17·2 |
| Liquor | 13·0 | 2·5 |
| Light Oils | 1·4 | 0·8 |
| Tar | 8·0 | 4·5 |
| Coke | 70·0 | 75·0 |

before the Second World War, several industries were almost entirely dependent upon coal tar for their existence, for example, those producing resins and plastics, dyestuffs, explosives, solvents, wood preservatives and disinfectants (Fig. 1). Since that time, the scene has been dominated by the growth of cheap petrochemical stocks, but now this position is changing again as crude oil becomes more expensive.

The first part of this chapter deals with the traditional coal tar and by-product industries. This is followed by a section on new uses for coal tar derivatives, particularly in the building and construction industries. Finally, there is a section on special carbons which can be made from coal.

## II. FORMATION OF BENZOLE AND TAR IN THE COKING PROCESS

The structure of coal and its behaviour on heating have already been discussed in Chapter 1. Of importance in this context are the groups of partly saturated rings and substituted, polynuclear aromatic structures which are 'loosely' connected in the coal substance. At its simplest, the coal 'molecule' can be thought of as being made up of several rings which are interconnected by various linkages to give a basic skeleton. Such structures degrade when heated in an inert atmosphere, as in the carbonization process, and breakdown occurs at the linkages in ways which are dependent upon time and temperature.

When coal is heated slowly, fragments break off progressively and at 450°C simple gases and primary tar products are given off. At low temperatures (e.g. below 500°C) the primary products contain appreciable quantities of naphthenes and paraffins, often with alkyl and hydroxyl substituents. As the temperature is increased, primary fragments must pass through a layer of hot coke and cracking occurs both there and on the walls of the oven, resulting in the conversion of the naphthenic structures to aromatics and removal of side chains. If the temperature is high enough, as in the production of blast furnace coke, then the tars produced consist mainly of aromatic hydrocarbons and higher boiling polynuclear compounds, and pitch formed by simultaneous polymerization of the unsaturated materials. The gases become progressively richer in hydrogen, and carbon is deposited as the temperature increases.

## III. PROCESSES

### Tar recovery

At present, the carbonization industry operates two main processes. The first produces metallurgical cokes at high temperatures (900-1100°C) for use in

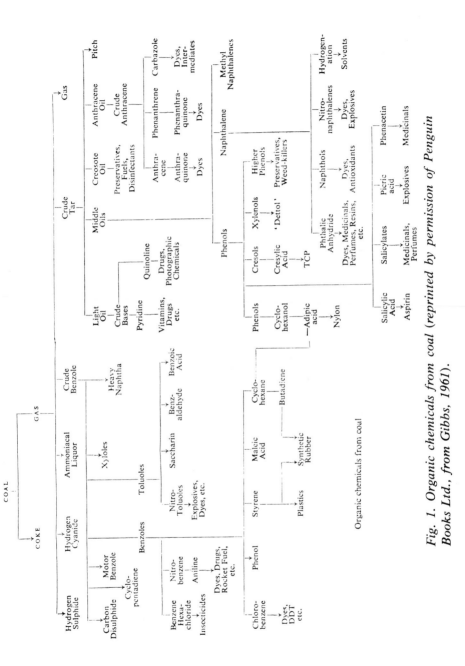

*Fig. 1. Organic chemicals from coal (reprinted by permission of Penguin Books Ltd., from Gibbs, 1961).*

blast furnaces and foundries. Separate batches, of about 15 tonnes of coal each, are charged into single slot ovens grouped together in batteries. The ovens are heated by burning product gas in flues between each oven. This is by far the most common method of producing coke.

The other method used, on a very much smaller scale, by the Coalite and Chemical Company, is a low temperature process in which coal charges of 0·3 t are carbonized in special iron retorts at 600°C to produce smokeless fuel for domestic use.

In both processes, the gas, tar and liquor vapours formed from the decomposing coal are taken into condensers. A typical system for tar condensation and by-product recovery is shown in Fig. 2.

The differences in yields of products and their compositions are determined mainly by the temperature and type of oven and, to a lesser extent, by the coals used. Figure 3 shows the yields of primary products and Fig. 4 gives the yields of tar fractions, both expressed against carbonization temperature. Broadly speaking, low temperature carbonization shows a lower gas yield, higher liquid yield with more tar acids, naphthenes and paraffins, but less pitch than high temperature tar. Tables II and III show analyses in more detail for high and low temperature tars.

Coke oven gas containing hydrogen, methane and carbon monoxide is used primarily to fuel the coke ovens and the excess is supplied to local industries. It is so rich in hydrogen that the possibility of separating it for use in chemical processing is currently being investigated in Germany.

The liquor is condensed separately as a solution of ammonia. It is then steam-stripped and treated with lime, before conversion into ammonium sulphate by sulphuric acid. Ammonium sulphate is mainly used as a fertilizer and to make fire-retardant mixtures. This outlet grew from impregnating wooden pit props with ammonium sulphate and has since been extended by incorporating better retardants such as phosphates, for treating paper, chipboards, hardboards, etc. The phosphate mixture is also used in dry powder fire extinguishers for all types of fires, including chemical fires.

Figure 5 shows a typical recovery plant for making benzole. In essence, it is a complicated system for stripping benzole from the wash oils in which it is collected.

## Treatment of Benzole

Most of the crude benzole is carried over with the gas and is recovered in a tower by countercurrent scrubbing with oil, usually creosote or petroleum-based gas oil. The benzole dissolves out of the gas and enriches the oil, which is distilled periodically to separate the crude benzole and wash oil for recycle, as in Fig. 5. A typical benzole analysis is shown in Table IV, together with an analysis for a refined product.

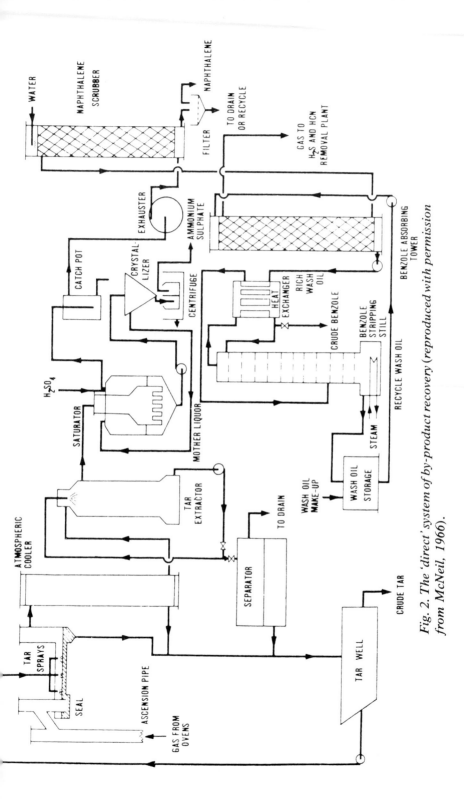

Fig. 2. The 'direct' system of by-product recovery (reproduced with permission from McNeil, 1966).

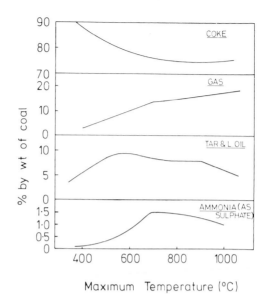

*Fig. 3. Carbonization yields of primary products versus temperature.*

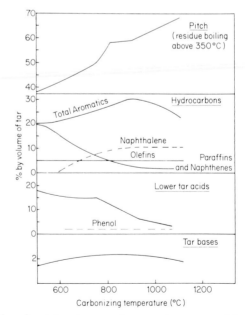

*Fig. 4. Effect of carbonizing temperature on tar composition.*

**Table II. Products of high temperature carbonization (900-1100°C).**

| Gas vol. % | | Light oil vol. % | | Tar vol. % | | Coke |
|---|---|---|---|---|---|---|
| $H_2$ | 50 | Benzene | 72 | BTX | 0·6 | |
| $CH_4$ + higher paraffins | 34 | Toluene | 13 | Phenols and cresols | 1·6 | |
| | | Xylene | 4 | | | |
| CO | 8 | | | Xylenols | 0·5 | |
| | | Alicyclics | 5 | | | |
| $CO_2$ | 3 | | | Other phenols | 1·0 | 1-2% Volatile matter |
| | | Aliphatics | 6 | | | |
| Other | 5 | | | Naphthalene | 8·9 | |
| | | | | Anthracene | 1·0 | |
| | | | | Other aromatics | 24·6 | |
| | | | | Tar bases | 1·8 | |
| | | | | Pitch | 60·0 | |
| Yield = 0·3m litres/tonne | | Yield = 14 litres | | Yield = 27-32 litres | | 0·7-0·8 tonne |

**Table III. Products of low temperature carbonization (400-750°C).**

| Gas vol. % | | Light oil vol. % | | Tar vol. % | | Coke |
|---|---|---|---|---|---|---|
| $H_2$ | 10 | Paraffins | 46 | BTX | 1·5 | |
| $CH_4$ + higher paraffins | 65 | Olefins | 16 | Phenol | 1·5 | |
| | | Cyclo-paraffins | 8 | Cresols | 4·5 | |
| CO | 5 | Cyclo-olefines | 9 | Xylenols | 7·0 | |
| $CO_2$ | 9 | Aromatics | 16 | Other phenols | 16·0 | 8-20% Volatile matter |
| Others | 11 | Others | 5 | Tar bases | 2·0 | |
| | | | | Naphtha | 3·5 | |
| | | | | Other aromatics | 38·0 | |
| | | | | Pitch | 26·0 | |
| Yield = 0·1m litres/tonne | | Yield = 11-16 litres | | Yield = 77-86 litres | | 0·75 tonne |

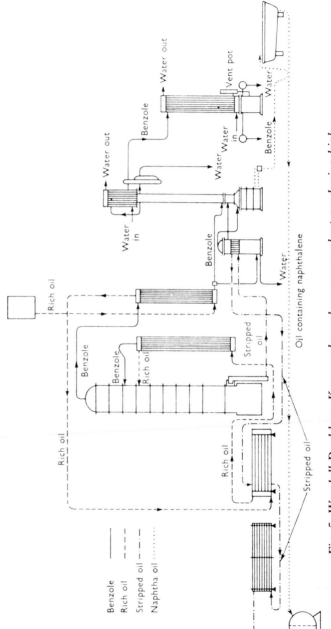

Fig. 5. *Woodall-Duckham-Koppers benzole recovery plant producing high-grade crude benzole (Woodall-Duckham Construction Co. Ltd.).*

Benzole ————
Rich oil — — —
Stripped oil — · — · —
Naphtha oil ············

Rich oil

Rich oil

Benzole

Benzole

Rich oil

Rich oil

Stripped oil

Stripped oil

Oil containing naphthalene

Water

Benzole

Benzole

Water

Water out

Water in

Water out

Water

Water
in

Water out

Benzole

Vent pot

Benzole

Water

*Table IV. Analyses on crude and refined benzole as in the Litol process (% weight).*

|                | Feed     | Product |
|----------------|----------|---------|
| Benzene        | 67·0     | 83·0    |
| Toluene        | 17·2     | 12·8    |
| Xylenes        | 4·6      | 0·8     |
| Ethyl benzene  | 0·9      | 1·1     |
| Styrene        | 1·4      | Nil     |
| Non-aromatics  | 1·4      | 0·4     |
| Thiophene      | 0·9      | Nil     |
| Indene         | 1·7      | Nil     |
| Biphenyl       | Nil      | 1·1     |
| Naphthalene    | 2·2      | Nil     |
| $C_9$ aromatics | 0·9      | Nil     |
| $CS_2$         | 272 ppm  | Nil     |

A process often used for this refinement is the Litol Process, shown in Fig. 6. Three separate treatments are involved:

(1) The conversion of sulphur compounds to hydrogen sulphide.
(2) The hydrocracking of non-aromatics to methane homologues.
(3) The dealkylation of part of the toluene and xylenes (cf. also Table IV).

The benzene-rich material is further purified by removal of olefines on passing through a bed of clay. The product is sold for conversion into cyclohexane and, later, into the Nylon 6 intermediate, caprolactam.

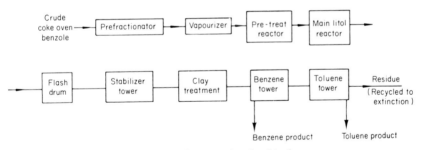

*Fig. 6. Principal stages in the Litol process.*

## Tar distillation

The recovered crude tar is distilled in a continuous pipe still to give the four standard fractions in order of ascending boiling ranges: light oil, naphthalene oil, creosote and anthracene oil plus pitch. A flow sheet for a continuous distillation plant is shown in Fig. 7. Crude tar is first dehydrated then fractionated in a distillation tower. Pitch and anthracene oil are separated in other columns. Table V gives the yields of various fractions.

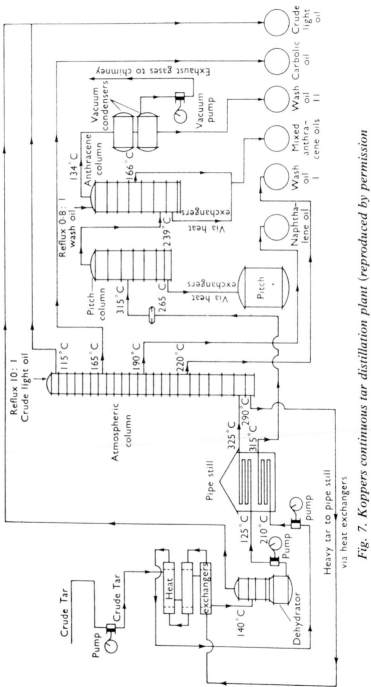

Fig. 7. Koppers continuous tar distillation plant (reproduced by permission of the Institution of Gas Engineers).

*Table V. Fractions obtained by distillation of high temperature tar.*

| Products | Boiling range °C | % Wt |
|---|---|---|
| Light Oil | < 195 | 1·0 |
| Naphthalene Oil | 195-230 | 12·0 |
| Creosote Oil | 230-300 | 6·0 |
| Anthracene Oil | >300 | 20·0 |
| Pitch | | 60·0 |
| Tar Acids | | 1·0 |

## IV. DISTILLATES

The light oil fraction contains benzene, tar acids and tar bases and resembles the crude benzole. It is sometimes added to benzole before refining.

The naphthalene oil fraction contains naphthalenes and a range of tar acids and tar bases. If present in sufficient quantity, as in low temperature tars, the tar acids and bases are extracted by washing successively with alkali and acid. Separated tar acids are regenerated and then distilled into fractions which are converted chemically into weed-killers, disinfectants, insecticides, plasticisers, anti-oxidation agents, mineral flotation agents, resins and even vitamins. Table VI shows the products made from Coalite tar acids. The neutral fraction usually contains about 30% naphthalene, which is removed either by crystallization and hot processing, continuous fractionation, or by crystallization and centrifugal washing. Formerly, pure naphthalene was used to make phthalic anhydride, an intermediate for plasticizers, polyesters and alkyd resins, but petroleum-based phthalic anhydride obtained via o-xylene has largely replaced this source.

The creosote fraction contains substituted naphthalenes, the higher boiling tar acids and base oils. The components are not usually separated and creosotes are used in bulk for timber preservation, for blending with pitch to make road tar, and as agricultural sprays. Creosote is also used as the wash oil for scrubbing benzole from the coke oven gas stream.

The anthracene oil fraction contains mainly polynuclear hydrocarbons, such as anthracene, phenanthrene and pyrene, and high boiling tar acids. At present, these oils find bulk uses similar to creosote, but they could be made into various resin intermediates for use in high temperature plastics. It is interesting that some anthracene is still purified for conversion to anthraquinone and dyestuffs, a process which has survived from the nineteenth century.

The pitch residue is used generally as a binder, e.g. to hold together the grist in electrodes for producing aluminium from bauxite (discussed later) and as a briquetting binder (see Chapter 3). It is also used in blends for road tar. Pitches are often heat-treated, blown with air or blended with polymers, to improve the properties.

*Table VI. Chemical products from the Coalite process.*

| Primary products | Chemicals extracted | Principal derivatives | Commercial outlets |
|---|---|---|---|
| Aqueous liquor | Catechol and homologues | Various mono, di and tri alkyl compounds. | Dyeing, pharmacy, oxidation inhibitors. Resins, adhesives, dyestuffs. |
| | Phenol | 98% Phenol mono, di and tri chloro compounds. | Resins, explosives, dyestuffs, insecticides, fungicides, plasticizers. |
| Bulked crude tar acids | Cresols Xylenols | Mono, di and tri alkyl compounds. | Various agricultural chemicals, anti-oxidants, inhibitors, antiseptics and disinfectants, selective weedkillers, pharmaceuticals. |
| | High boiling tar acids | Various grades—some chlorinated. | Disinfectants and high grade antiseptics. |

## V. NEW USES FOR COAL TAR MATERIALS

In spite of the loss of many traditional markets to petroleum, there are several outlets for coal tar products. One market which is expanding is in proprietary products related to building. The excellent waterproofing properties and corrosion resistance of pitch have led to the development of a pitch polymer sheet, commercially known as Hyload. This was originally intended for use as a damp-proof membrane in high rise buildings. The basic formula has since been modified many times, e.g. as a roof sheeting and for bridge decking damp-proof coursing. A remedial waterproof paint (Synthaprufe) has long been used for treating walls and small roofs. This basic idea has been extended to anti-corrosion treatments and various tar-based paints and coal/pitch/epoxy compositions have been produced.

In entering the housing market with Hyload, it soon became clear that a need existed to provide a dampcourse for the many houses already built before dampcoursing was commonplace. A treatment was soon developed, based on a solution of silicone resin in coal tar naphtha. An *in situ* dampcourse is formed by injecting the solution into the walls of houses, where the naphtha evaporates and the silicone gels into a waterproof layer. Moreover, a similar solution (Synthasil) can be sprayed or brushed onto porous masonry to weatherproof outside walls. Well over 100 000 houses have been treated by these methods in the last three years. In both treatments, a continuous film of silicone resin, impervious to water, is left after the evaporation of the solvent naphtha (Fig. 8). Experience in field tests with silicones also showed the need for timber preservation and so led to the production of solutions based on pentachlorophenol dissolved in light oil fractions for eradicating and preventing dry and wet rot.

As part of the development programme on building materials, heat loss measurements were made on dry, damp and very wet houses, thus leading to a study of insulants for housing and industry. Many existing insulation boards in roofs, partitions, etc., have a number of drawbacks, such as being flammable, dripping, giving off highly toxic fumes in fires or becoming slowly waterlogged and being relatively poor in insulation efficiency. A type of insulation board based on phenol formaldehyde resin foam (see Fig. 9) has been evolved in which these defects are largely eliminated. Mention has already been made of other fire retardants made from coal-based materials. Their application in building materials, such as chipboard, is especially important.

Although many traditional markets for coal tar chemicals have been taken over by the petrochemical industry, the position is changing again with rising oil prices. Recently, a series of new resins based on coal-derived naphthalene and toluene with formaldehyde was launched commercially. These resins are

*Fig. 8. Silicone injection dampcourse.*

*Fig. 9. Phenol formaldehyde foam insulation boards being laid.*

used as modifiers in the rubber industry, as additives, as extenders to existing, more expensive resins and as protective and insulating materials in electrical installations.

In addition, a process for extracting light oil with *N*-methyl pyrollidone to provide special aliphatic and aromatic solvents has been introduced recently.

## VI. SPECIAL PRODUCTS

### Special carbons

In Chapter 8, a solvent extraction process which produces an ash-free 'coal solution' is described. This primary coal extract is a versatile starting material for a wide range of products, among them some special and interesting carbons.

If coal extract is coked in a delayed coker at 500°C, then a special coke is produced which is a valuable carbon for the steel and aluminium industries. This coke is often termed 'green coke' or 'needle coke', the variety preferred for the manufacture of electrodes for the arc steel furnace.

For this purpose, green coke is calcined at 1300°C in order to make coke 'grist' before being mixed with about 25% hot pitch. The hot mixture is extruded into rods which are baked to 1000°C very slowly in order to minimize physical deformation. To improve mechanical strength and electrical conductivity, the baked rods are often impregnated with additional pitch and baked again. The rods are finally heated in an electric furnace to 2500°C when graphitization occurs and electrical conductivity increases greatly. Trial batches of coke from coal extract have been made up into extruded rods of 300 mm diameter (Fig. 10) and tests in an arc furnace are now in progress. The structure and properties of graphite electrodes have been dealt with already in Chapter 4.

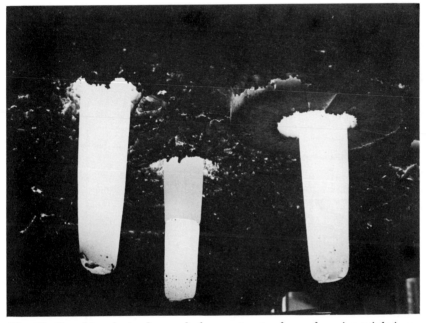

*Fig. 10. Graphite electrodes made from extract coke undergoing trials in an arc steel furnace.*

A second market for coal extract coke that deserves attention is in the manufacture of baked carbons (not graphitized) for use as anodes in making aluminium. This is a much larger market than that for arc steel electrodes as just over 0·5 tonne of carbon is consumed in producing every tonne of aluminium from bauxite, compared with 5-6 kg of graphite consumed per tonne of arc steel. At present, both markets are supplied exclusively from petroleum refinery products. However, this is not likely to continue as crude oil supplies decline.

## Active carbon

Another completely different form of carbon can be made from coal in which
it is the active surface which is all important. Such carbons are used as
absorbents and are generally known as active carbons and certain coals are
good starting materials.

A process for the activation of anthracite has been developed to the
commercial stage and a plant is in operation at Coedely in South Wales. The
process depends on partial gasification in steam and makes use of fluidized
bed technology. A flow sheet is shown in Fig. 11.

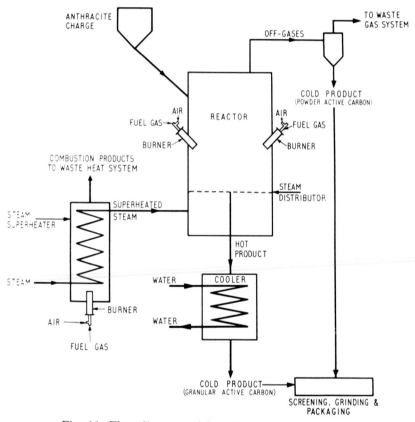

*Fig. 11. Flow diagram of Coedely active carbon plant.*

Anthracite has been considered as a feedstock for new methods of active
carbon manufacture by both American and Russian workers and it may also
be blended with other coals for use in established processes which incorporate
a briquetting or an extrusion stage. The Coedely plant, however, is the only

one known to be operating commercially, solely on an anthracite feedstock. Anthracite gives a high yield and the fluidized bed process gives rapid and uniform treatment of the granular product.

The activation process is based on the removal of carbon atoms, as gaseous oxides, by the reaction of carbon and steam, leaving behind a system of interconnected internal pores. Carbon-steam reactions are the same as those used in coal gasification processes (Chapter 7) but in activation the minimum conversion to gas is aimed at, consistent with adequate burn-off for the required pore structure. The plant operates batchwise and by varying the extent of carbon burn-off, the activity of the products can be controlled. The main operating parameters are reaction time, bed temperature and steam flow rate. With progressive burn-off, the pore volume and mean pore size of the product increase, but at the expense of a reduced yield.

To obtain a reasonable rate of reaction without excessive surface burn-off, the process temperature is normally within the range 750-1000°C. Typical analyses of the feedstock and a granular carbon are shown in Table VII.

Volatile matter, mainly hydrogen, is given off by the anthracite as it is heated to the reaction temperature. During this stage, steam is present and this may be important in preventing pore closure which can occur at high temperatures.

There are two products, the granular carbons forming the fluidized bed and the fine carbon carried out of the reactor by the off-gases.

The method of supplying the reaction heat is unusual. The plant is sited at a coking works and near a colliery, and both coke oven gas and mine drainage methane are available. Gas burners which can use either fuel are immersed in the bed; the carbon dioxide and water produced by the combustion are available as reactants along with the fluidizing steam. To prevent free oxygen entering the bed, the burners are operated near stoichiometric conditions.

Several methods are available to determine the adsorption characteristics of the products. For routine quality control, the amounts of model

Table VII. Analyses on active carbon and anthracite.

| | | Anthracite | Carbon |
|---|---|---|---|
| *Proximate analysis* (wt %) | | | |
| VM | (daf) | 6·0 | 2·8 |
| Ash | (dry basis) | 3·1 | 5·1 |
| *Ultimate analysis* (wt % dmmf) | | | |
| C | | 93·5 | 96·3 |
| H | | 3·2 | 0·7 |
| O | | 1·0 | 1·9 |
| N | | 1·3 | 0·7 |
| S (total) (dry basis) | | 1·1 | 0·6 |

adsorbates, such as methylene blue, iodine, carbon tetrachloride, benzene, etc., adsorbed at equilibrium are determined. For a more fundamental characterization, the total pore volume and the distribution of pore volume in respect of pore entry size are determined using density measurements in different fluids, mercury porosimetry and low temperature nitrogen adsorption. Application of these different techniques has shown that the active carbons from anthracite have a considerable pore volume and an abundance of micropores (see Fig. 12).

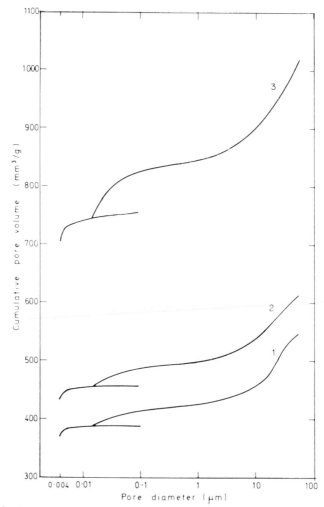

*Fig. 12. Pore diameter/pore volume relationship for three active carbons made from anthracite.*

*1 and 2: commercial grades*

    *3: special high surface carbon.*

The largest markets are concerned with water and effluent treatment. Powdered carbon is used by several UK Water Authorities to control taste and odour, and granular grades are also sold for this use in other European countries. A related use is the dechlorination of water on ships, oil rigs and for breweries. Granular carbon finds applications in a variety of effluent problems, e.g. removal of detergents from vehicle washing plants and of biocides in the manufacture of pesticides. Purification of a process liquid before recycling is another use for powdered carbon, e.g. of dry cleaning fluid and of electroplating solutions. Powdered carbon is also used as a catalyst support. Vapour phase applications of granular and pelleted forms of active carbon cover air conditioning, use in cooker hoods, adsorption of petrol vapour in road vehicles and solvent recovery. In the last example cited, regeneration of the granular active carbon is accomplished by stripping with steam at comparatively moderate temperatures. In many cases, however, where granular active carbon is used, regeneration has to be carried out at temperatures approaching those of the initial activation process, and generally without recovery of the adsorbates.

With the increasing emphasis on making the environment cleaner, active carbons are likely to be needed on an increasing scale.

## Carbon Fibres

At first sight, it may seem surprising that a complex substance like coal can be used to produce a high purity product such as carbon fibre. This can be brought about by first making a coal extract (i.e. a solution of coal in a suitable solvent as mentioned above) as an intermediate which removes the inorganic matter (or ash) from the coal. The process for converting the coal extract into carbon fibre has four stages: spinning, oxidation, carbonization and strain graphitization.

In spinning, the coal extract is extruded from the melt under pressure and drawn down to diameters in the range 10-30 $\mu$m while still soft; the fine filaments cool rapidly and are wound onto a bobbin to make a package of raw, continuous, multifilament fibre.

Oxidation and carbonization of the filaments are carried out continuously in horizontal furnaces (see Fig. 13). Oxidation consists of heating in air or in oxygen to temperatures over the range 100-300°C. This renders the filaments sufficiently infusible to prevent them from softening and sticking together during carbonization when the filaments are heated in an inert atmosphere to a temperature of 1000°C. The final step of strain graphitization is carried out continuously in a high-temperature induction furnace at 2700°C. Strain graphitization orientates the graphitic structure and increases both the modulus and the tensile strength of the fibre by factors dependent on the degree of imposed strain. It is possible to exceed 100% strain, and a modulus

*Fig. 13. Experimental preparation of carbon fibre.*

of 600 GN m$^{-2}$ with strength of 3 GN m$^{-2}$ can be obtained. The mechanical properties of typical carbonized and strain graphitized fibres are given in Table VIII.

Structural information having a direct bearing on the mechanical properties of the fibre has been obtained by a number of techniques. The carbonized fibre gives rise to a diffuse X-ray diffraction pattern indicative of the presence of only short range order, while for the high modulus, strain

*Table VIII. Properties of typical carbon fibres made from coal extract.*

| Property | | Carbonized fibre | Strain graphitized fibre |
|---|---|---|---|
| Tensile strength | $(\sigma)$ | 0·8  GN m$^{-2}$ | 2·4  GN m$^{-2}$ |
| Young's modulus | $(E)$ | 45    GN m$^{-2}$ | 400    GN m$^{-2}$ |
| Elongation at break | | $\sim 2\%$ | $\sim 0.5\%$ |
| Specific gravity | $(sg)$ | 1·65 | 1·80 |
| Specific strength | $\dfrac{(\sigma)}{(sg)}$ | 0·5  GN m$^{-2}$ | 1·3  GN m$^{-2}$ |
| Specific modulus | $\dfrac{(E)}{(sg)}$ | 27    GN m$^{-2}$ | 220    GN m$^{-2}$ |
| Electrical resistivity | | 80 $\mu\,\Omega$ m | 10 $\mu\,\Omega$ m |

graphitized fibre all the reflections are much sharper and there is a distinct preferred orientation of the 001 planes along the filament axis. This is supported by the results of high resolution electron microscopy (see Chapter 5).

Photomicrographs of broken ends of the carbonized fibre show a typical conchoidal fracture of an isotropic material, whereas those for the strain graphitized fibre (see Fig. 14) show an anisotropic structure, which in many respects resembles the fibrillar structure observed in fibre made from polyacrylonitrile (PAN).

Thus, it has been demonstrated experimentally that coal can be transformed into top quality carbon fibre. The strain graphitized fibre has properties equal to those of the best commercially available PAN fibre. The coal-based route should, if carried out on a sufficiently large scale, have a cost advantage over the PAN route because of the relatively low cost of the extract and the high carbon yield. However, in spite of its promise as a new material, growth in markets has been slow and world production is still only a few hundred tonnes per annum. So far, this has not warranted further development of the process route from coal.

## VII. CONCLUSION

There is a large industry based on the by-products obtained when coal is carbonized. In the UK, about one million tonnes per annum of tar products are supplied to traditional markets, e.g. solvents, timber preservatives, fertilizers, briquetting binders, electrode binders and road dressings. In addition, a range of useful products is now made for the building trade, including waterproofing sheets and solutions, insulation materials, resins and

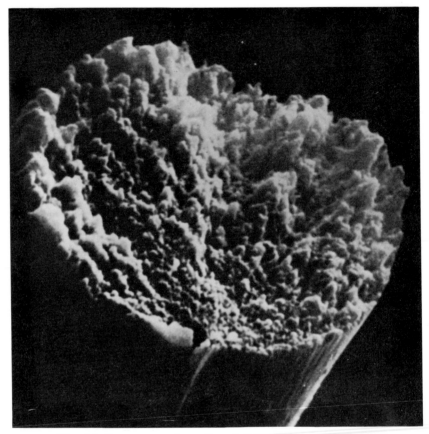

*Fig. 14. Fracture surface of strain graphitized carbon fibre.*

fire retardants. Recently, there has been a return to supplying chemicals, e.g. via the Litol process, and for making specialized resins.

Special carbons, e.g. cokes for graphite electrodes and carbon fibres have already been produced in experimental equipment, while active carbons made from coal are available commercially.

## REFERENCES

Gibbs, F. W. (1961). "Organic Chemistry Today." pp. 96-97. Penguin Books Ltd., London.
McNeil, D. (1966). "Coal Carbonisation Products." Pergamon Press Ltd., Oxford.

# Subject Index